A BRIDGE
TO HIGHER
MATHEMATICS

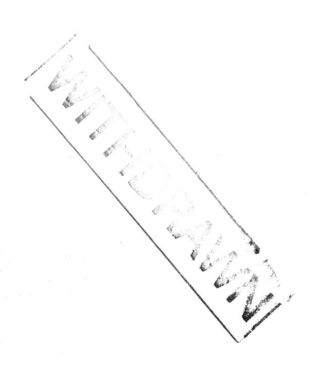

TEXTBOOKS in MATHEMATICS

Series Editors: Al Boggess and Ken Rosen

PUBLISHED TITLES CONTINUED

TEXTBOOKS in MATHEMATICS

A BRIDGE TO HIGHER MATHEMATICS

Valentin Deaconu
University of Nevada
Reno, USA

Donald C. Pfaff
University of Nevada
Reno, USA

CRC Press
Taylor & Francis Group
Boca Raton London New York

CRC Press is an imprint of the
Taylor & Francis Group an **informa** business
A CHAPMAN & HALL BOOK

CRC Press
Taylor & Francis Group
6000 Broken Sound Parkway NW, Suite 300
Boca Raton, FL 33487-2742

© 2017 by Taylor & Francis Group, LLC
CRC Press is an imprint of Taylor & Francis Group, an Informa business

No claim to original U.S. Government works

Printed on acid-free paper
VersionDate:20161102

International Standard Book Number-13: 978-1-4987-7525-0 (Paperback)

Visit the Taylor & Francis Web site at
http://www.taylorandfrancis.com

and the CRC Press Web site at
http://www.crcpress.com

Printed and bound in the United States of America by Sheridan

Contents

Contents

Preface

To the student

There are college students who have done quite well in mathematics up to and including calculus, but who find that their first encounter with upper division math is a somewhat traumatic experience. The main reason for these difficulties may be the fact that excelling in the purely computational aspects of math like the plug and chug method is not sufficient for writing arguments down and proving theorems in classes like Real Analysis, Abstract Algebra or Topology. Writing proofs is a skill you acquire with lots of practice, similar to integration by substitution, computing a derivative using the differentiation rules, or factoring a polynomial to find the roots.

In order to be able to read this book, we assume that you have knowledge about basic algebra, divisibility of integers, trigonometry, some plane geometry and calculus. Many of you may have seen these subjects in high school or in the first two years of college. In particular, we use the notation $\mathbb{N}$ for the set of natural numbers $0, 1, 2, 3, ...$; $\mathbb{Z}$ for the set of integers $..., -2, -1, 0, 1, 2, ...$; $\mathbb{Q}$ for the set of rationals (quotients of integers p/q excluding $q = 0$) and $\mathbb{R}$ for the set of reals (rationals plus numbers like $\sqrt{2}, \pi$ and decimal numbers with nonrepeating decimals), which appear in many calculus textbooks. We will also use the notation $f : \mathbb{R} \to \mathbb{R}$ for a function defined on $\mathbb{R}$, which takes real values. If the domain of the function is just an interval I, we will write $f : I \to \mathbb{R}$.

Throughout this book, we will emphasize the necessity of proving facts, and we will introduce new symbols and concepts necessary for a transition to proof-based courses. Sometimes, you will have the feeling that we will do things all over again, but from a different point of view. For example, you may know that if the square of an integer n is even, then n must be even. But how do you prove that rigorously? You will learn the language of axioms and theorems and you will write convincing and cogent proofs using quantifiers. You will solve many exercises and encounter some challenging problems.

It is common for a college student to stand in awe of the professor and perhaps to regard a textbook writer as a person with some kind of superhuman knowledge. What is often overlooked is the fact that we professors were once students ourselves and many of us thought that we would never attain the lofty heights of understanding that our teachers displayed so casually during the course of their lectures.

Let it be known right now that a substantial number of readers of this book are much brighter and faster than we ever were. If we have an advantage over them, it is merely due to experience, not to innate ability. The purpose in writing this book is to share some things that we have learned over the years; ideas, examples and techniques we have found helpful and which we believe can help the students to progress faster and further along the road to deeper mathematical understanding.

This process is not easy and many times you will have to ask yourself: did I understand the definition of this concept? What can I assume to be known? What exactly do I have to show? What method or strategy should I use? Only lots of practice and patience will help you to find good answers to these questions. The only way to understand mathematics is by doing mathematics. You cannot learn to play the piano by watching somebody playing the piano; mathematics is not a spectator sport.

To the instructor

The material covered in the first nine chapters of this book can be used for a course in the first semester of the junior year for math majors. We begin our journey with elements of logic and techniques of proof, then we continue with elementary set theory, relations and functions, giving many examples, some of them contained in the exercises. There is no claim of self-contained axiomatic introduction to logic or set theory. We discuss the Peano axioms for positive integers and for natural numbers, in particular mathematical induction and other forms of induction. Assuming that we are familiar with the integers, we discuss divisibility tests, the Euclidean algorithm for finding the greatest common divisor and the fundamental theorem of arithmetic. We continue with the notions of finite and infinite sets, cardinality of sets and then we discuss counting techniques and combinatorics, illustrating more techniques of proof. Some advanced topics, like Zorn's lemma and the axiom of choice are marked with an asterisk and they could be the subject of a project. Sometimes our discussions and proofs are incomplete, and we direct the reader to consult other books. Also, we included some more challenging problems, marked with a star. All these materials are optional, depending on the instructor and the goals of the reader.

For the interested students, we include in the book a rigorous construction of the set of integers $\mathbb{Z}$, by using an equivalence relation on pairs of natural numbers. Similarly, the set of rational numbers $\mathbb{Q}$ is introduced using an equivalence relation on pairs of integers $\langle m, n \rangle$, where $n \neq 0$. The set of real numbers $\mathbb{R}$ is constructed from the rationals by using two methods: Dedekind cuts and Cauchy sequences. We also construct the set of complex numbers $\mathbb{C}$, as pairs of real numbers, and talk about the trigonometric form and some connections with plane geometry.

A word of caution: different books may use different notation for the same notion. Also, the same symbol may have different meanings in different contexts. This is sometimes for historical reasons, sometimes from laziness, but many times because it is difficult to invent new and meaningful signs that will please everybody. For example, (a, b) is often used to denote an open interval in $\mathbb{R}$, but also to denote an ordered pair or a vector in the plane; $[a, b]$ means a closed interval, but other times means an equivalence class; $\wedge$ is a logical connector, but is also used to denote the greatest lower bound, etc. We chose to use angular brackets $\langle a, b \rangle$ for ordered pairs, $A \subseteq B$ for inclusion of sets, and $A \subset B$ for strict inclusion.

We included hints and solutions to some of the exercises. Keep in mind that sometimes there are different proofs of the same result. In principle, we do not believe in giving all the answers in the back of the book. So many times the students try, by reverse engineering, to figure out the solution from the answer. The students should be instructed how to break up the objective into manageable parts, read the definitions carefully and understand the examples before trying to solve the exercises.

One of us (Pfaff) has spent a substantial part of his career teaching classes for those who intend to be secondary school teachers. This experience tells us that the gap between elementary and advanced math is, for most persons, more of a yawning chasm than a gap. The problem seems to be that very few really understand the mechanics and processes that constitute a correct mathematical proof. To progress toward mathematical maturity, it is necessary to be trained in two aspects: the ability to read and understand a proof and the ability to write a proof. It is our goal to make a succinct and smooth transition to this kind of maturity.

Some of the individual facts and examples presented in this book may be already familiar to the student. This is deliberate. By dealing with material that they have already seen in high school and in the first two years of college, the students will be in a better position to concentrate on the underlying thought processes and practice, in a variety of applications, the many theorem-proving techniques that should be part of every mathematician's arsenal.

There will be, of course, numerous new concepts and results. We have in mind all the tools that will be necessary to make the transition from lower-division courses like Calculus, Differential Equations and Linear Algebra to upper-division classes like Abstract Algebra, Real and Complex Analysis and Topology to name a few.

The emphasis in this book, as you may have guessed, will be on proof. But we intend to do more than simply prove a bunch of theorems. All mathematically literate persons should be conversant with the basics of math, including a knowledge of logic, sets, functions, relations, and the different kinds of numbers and their properties. So, we intend to cover the fundamentals of abstract mathematics, but with special attention paid to its logical structure and especially with an emphasis on how the theorems are proved.

Aknowledgments. We thank the editor and the reviewer at CRC Press who helped us improve the exposition and the content of this book.

1

Elements of logic

Logic is the systematic study of the form of valid arguments. Valid arguments are important not just in mathematics, but also in computer science, artificial intelligence or in everyday life.

In this chapter we introduce mathematical statements and logical operations between them like negation, disjunction and conjunction. We construct truth tables and determine when two statements are equivalent, in preparation for making meaningful judgments and writing correct proofs. We introduce new symbols, like the universal quantifier $\forall$ and the existential quantifier $\exists$, and learn how to negate statements involving quantifiers. Quantifiers will be used throughout the book.

1.1   True and false statements

We can use words and symbols to make meaningful sentences, also called statements. For example:

a) Mary snores.
b) A healthy warthog has four legs.
c) $2 + 3 = 5$.
d) $x + 5 = 7$.
e) $\int_0^\pi \sin x \, dx = 2$.
f) $\forall x \in \mathbb{R} \; \exists \, y \in \mathbb{R}$ such that $y^2 = x$.
g) $x/x = 1$.
h) $3 \in [1, 2)$.
i) Dr. Pfaff is the president of the United States.

Some of these statements have eccentric formats, using symbols that you may not have seen before. By the way, $\forall$ means for all, $\in$ means belongs to (or is an element of) and $\exists$ means there exists. The symbol $\forall$ is also called the *universal quantifier*, and the symbol $\exists$ is called the *existential quantifier*.

The statements could be true, like statements b, c, e; could be false, like statements f, h, i; or could be neither true nor false, like statements a, d, g.

1

The last possibility occurs because we don't have enough information. For statement a, we could ask: which Mary are we talking about? Is she snoring now or in general? For statement d, do we know that $x = 2$? For statement g, do we know x to be a nonzero number? We don't, and these kind of ambiguous statements (neither true nor false) also appear in real life.

To make things easier, let's agree that a *proposition* is a statement which is either true or false. Each proposition has a truth value denoted T for true or F for false. All statements b, c, e, f, h, i are propositions, but a, d, g are not. We can modify statements d and g as

d') $\forall x : x + 5 = 7$ and

g') $\exists x : x/x = 1$,

which become propositions. Of course, the proposition d' is false since, for example, $1 + 5 \neq 7$ and the proposition g' is true because we can take $x = 2$ and $2/2 = 1$.

1.2   Logical connectives and truth tables

It is important to understand the meaning of key words that will be used throughout mathematics. Primary words which must be clarified are the logical terms "not", "and", "or", "if...then", and "if and only if". The word "and" is used to combine two sentences to form a new sentence which is always different from the original ones. For example, combining sentences c and h above we get

$$2 + 3 = 5 \text{ and } 3 \in [1, 2),$$

which is false, since 3 does not belong to the interval $[1, 2)$. The meaning of this compound sentence can be determined in a straightforward way from the meanings of its component parts. Similar remarks hold for the other basic logical terms, also called connectives. We will think of the above basic logical terms as operations on sentences. Though there are linguistic conventions which dictate the proper form of a correctly constructed sentence, we will find it convenient to write all our compound sentences in a manner reminiscent of algebra and arithmetic. Thus, irrespective of where the word "not" should appear in a sentence, in order to placate the grammarians, we will write it sometimes at the beginning. For example, though a grammar book would tell us that "not Dr. Pfaff is the president of the United States" is improper usage, and that the correct way is to write "Dr. Pfaff is not the president of the United States", it will prove easier for us to work with the first form. We regard the two as the same for our purposes.

We will use the following symbolic notation.

Definition 1.1. *The* negation *of a statement P is denoted by $\neg P$, verbalized as "it is not the case that P" or just "not P".*

The operation of negation always reverses the truth value of a sentence. We summarize this in this truth table:

P	$\neg P$
T	F
F	T

Recall that T stands for true and F for false.

Definition 1.2. *We write $P \wedge Q$ for "P and Q", the conjunction of two sentences. This is true precisely when both of the constituent parts are correct, but false otherwise.*

This is in line with normal everyday usage of the word "and". Nobody would deny that I am telling the truth if I say

$$(2 + 2 = 4) \wedge (3 + 3 = 6),$$

nor would anyone hesitate to call me a liar if I boldly announced that

$$(2 + 2 = 4) \wedge (3 + 3 = 5).$$

Here is the truth table for the conjunction $P \wedge Q$:

P	Q	$P \wedge Q$
T	T	T
T	F	F
F	T	F
F	F	F

Definition 1.3. *We write $P \vee Q$ for "P or Q", the disjunction of two statements.*

There is a bit of a surprise when we consider the mathematical usage of the word "or". This is because it is often used in ordinary language to mean the same as "either ... or", excluding the possibility of two things being true at the same time. Many restaurants will offer you soup or salad with the main course, assuming that you cannot have both soup and salad. In mathematics, however, we use the word "or" in the inclusive sense of "at least one, possibly both". Thus all three statements

$$(2 + 2 = 4) \vee (3 + 3 = 5)$$

$$(2 + 2 = 5) \vee (3 + 3 = 6)$$

$$(2 + 2 = 4) \vee (3 + 3 = 6)$$

are true. The only "or" statement that is false is the one with both component parts false, as for example

$$(2 + 2 = 5) \vee (3 + 3 = 5).$$

In ordinary conversation, if I say "I will take you to the movies or buy you a candy bar", you would be probably satisfied if I did either one, but I'm sure you wouldn't call me a liar if I did both. That is an illustration of the proper usage of "or" in the mathematical sense. The truth table of the disjunction $P \vee Q$ is as follows:

P	Q	$P \vee Q$
T	T	T
T	F	T
F	T	T
F	F	F

You may detect a note of arbitrariness in our cavalier description of how the word "or" is to be used. There are, however, good reasons for this choice, and most mathematicians use this interpretation in textbooks and journal articles. Also, this particular way of employing the disjunction allows for some nice relationships between the logical connectives, much like the basic identities and laws of algebra that you may be familiar from high school.

To be fair, we mention that some people work with the logical operation *exclusive or*, denoted as $\underline{\vee}$ or $\oplus$ in a truth table:

P	Q	$P \oplus Q$
T	T	F
T	F	T
F	T	T
F	F	F

Notice that in this case, $P \oplus Q$ is true precisely when only one of the components is true and the other is false.

Definition 1.4. *A sentence of the form "if P then Q" is written symbolically $P \Rightarrow Q$ and it is called a* conditional *or* implication. *We can also read P implies Q, P only if Q, P is sufficient for Q, or Q is necessary for P. Some authors use the notation $P \to Q$. The sentence P is called the* hypothesis *or the* antecedent, *and Q is the* conclusion *or the* consequent.

An "if ... then" statement is more precisely defined when used in a logical context than when casually bandied about in ordinary speech. For our purposes, it is most important to understand that *any* two statements can be joined together to form a conditional; the individual parts can be true or false.

We will say that a conditional is false when the antecedent is true and the consequent is false. In all other situations, the conditional is taken to be true. Thus a conditional is understood to be true whenever the sentence following "if" and preceding "then" is false. As this convention may shock your tender sensibilities, we will try to motivate our reasons for choosing it by relating to some examples.

Perhaps it will help if you think of a statement of the form $P \Rightarrow Q$ as a promise with a condition (in fact, this is why such statements are called *conditionals*). The promised end need not be true unless the condition is met. Suppose I promise that you will get an A in the class IF you have an average which is 90% or greater. If I am not lying, then you will certainly expect an A if your average is 92.3%. You will, of course, be upset and complain if you breeze through with a 90% average and I give you a B. And you will be justified in your reaction. But if your average is 89.7%, I can give you any grade I want without breaking my promise. The question of what will occur for an average under 90% is simply not addressed by the promise as stated. To be more specific, suppose P is the statement "your average is 90% or better" and Q represents "your grade is A". The promise is symbolized as $P \Rightarrow Q$. Consider the four possible outcomes at semester's end:

1. Your average is 92.3% and your grade is A.
2. Your average is 92.3% and your grade is B.
3. Your average is 89.7% and your grade is A.
4. Your average is 89.7% and your grade is B.

In the context of my promise, the only blatant lie arises from situation number 2. I have kept my promise in all three of the other cases. When you fail to fulfill your part in the bargain, as when your average is 89.7%, you may be resigned to the fact that you will get a B, but I do not suddenly become a liar if, because of generosity or because I'm such a swell guy, I choose to give you the A.

Comedians have known about and used the mathematical interpretation of a conditional statement for a long time. Here it is:

"If you had two million dollars, would you give me one million?"
"Of course!"
"If you had two thousand dollars, would you give me one thousand?"
"Certainly!"
"If you had twenty dollars, would you give me ten?"
"No way!"
"Why not?"
"Because I have twenty dollars!"

The whole point is the idea that, without lying, one can promise *anything* in a conditional fashion, provided that the condition is not fulfilled. An important point is that the statement $P \Rightarrow Q$ does NOT guarantee anything about the truth of P or Q individually. The truth of a conditional merely expresses a connection between a hypothesis and a conclusion. Even if the conditional is true, you know that Q is correct ONLY after you have determined that P is correct.

Here is another example to illustrate that a false statement implies anything: let's prove both

(1) if $2 + 2 = 5$, then $3 = 0$,   and (2) if $2 + 2 = 5$, then $3 = 3$.

We start with the antecedent and, applying valid algebraic principles, we try to reach the conclusion. Let's start with (1): if $2 + 2 = 5$, then $4 = 5$. By subtracting 5 from both sides, we get $-1 = 0$. Multiplying both sides with -3, we get $3 = 0$. What do you think? Did we prove that $3 = 0$? Of course not. We proved it from a false assumption. The entire statement (1) must be regarded as true.

For (2), we have the hypothesis $2+2 = 5$. Multiply both sides by 0 and add 3 to both sides. We obtain a valid result $3 = 3$ from an erroneous assumption.

To summarize, the truth table for the conditional $P \Rightarrow Q$ is as follows:

P	Q	$P \Rightarrow Q$
T	T	T
T	F	F
F	T	T
F	F	T

Definition 1.5. *We write $P \Leftrightarrow Q$ for "P if and only if Q", and call it a biconditional or equivalence. We can also read P is equivalent to Q or P is necessary and sufficient for Q. Some people use the notation $P \leftrightarrow Q$. The expression "if and only if" is often abbreviated as "iff".*

A biconditional is used when we intend to express the idea that two statements surrounding it say the same thing, albeit in different ways. Since, at this stage, we are concerned only with truth and falsity, we agree that the statement $P \Leftrightarrow Q$ asserts that P and Q are both true or both false. As usual, we must understand that merely stating a biconditional does not make it true. Also, we do not make any a priori restrictions on the kinds of sentences which may be joined by the symbol $\Leftrightarrow$. There may or may not be a perceivable relation between the component parts of such a sentence. For our purposes, the decision as to truth or falseness of the whole is determined solely by an examination of the constituent parts. Thus the statement

$$(2 + 2 = 4) \Leftrightarrow (7 \text{ divides } 1001)$$

is true, never mind that you see no rhyme or reason for putting the two parts together. When two true sentences are combined by using the words "if and only if", the result is understood to be true. What do you think about the following sentence? True or false?

$$(7 < 5) \Leftrightarrow (\text{Don Pfaff played Han Solo in } \textit{Star Wars}).$$

Both component parts are false. Thus they convey the same information, even though that information is wrong in both cases. The entire statement is true. It asserts nothing about whether the individual parts are correct, only that they are equivalent with respect to their truth values.

The sentence

$$(2 + 2 = 4) \Leftrightarrow (1 = 0)$$

is false, since the parts do not have the same truth value. A biconditional is false if one of the component parts is right and the other is wrong. We have the following truth table for $P \Leftrightarrow Q$:

P	Q	$P \Leftrightarrow Q$
T	T	T
T	F	F
F	T	F
F	F	T

1.3 Logical equivalence

A truth table will show us that $P \Rightarrow Q$ has the same truth values in all cases as $(\neg P) \vee Q$. We say that $P \Rightarrow Q$ is logically equivalent with $(\neg P) \vee Q$. It is also logically equivalent to $(\neg Q) \Rightarrow (\neg P)$. Indeed, it suffices to look at the truth table:

P	Q	$\neg P$	$\neg Q$	$(\neg P) \vee Q$	$(\neg Q) \Rightarrow (\neg P)$	$P \Rightarrow Q$
T	T	F	F	T	T	T
T	F	F	T	F	F	F
F	T	T	F	T	T	T
F	F	T	T	T	T	T

Definition 1.6. *The statement* $(\neg Q) \Rightarrow (\neg P)$ *is called the* contrapositive *of* $P \Rightarrow Q$. *The statement* $Q \Rightarrow P$ *is called the* converse *of* $P \Rightarrow Q$.

A truth table shows that $P \Leftrightarrow Q$ is logically equivalent to $(P \Rightarrow Q) \wedge (Q \Rightarrow P)$.

The converse of a true statement is not necessarily true. For example, the converse of the statement "If n is divisible by 4, then n is divisible by 2" (true) is "If n is divisible by 2, then n is divisible by 4" (false). The contrapositive and the converse of a statement will be used in the next chapter, when we talk about several techniques of proof.

Definition 1.7. *Let A be a proposition formed from propositions $P, Q, R, ...$ using the logical connectives. The proposition A is called a* tautology *if A is true for every assignment of truth values to $P, Q, R, ...$ The proposition A is called a* contradiction *if A is false for every assignment of truth values to $P, Q, R, ...$*

For example, $P \wedge Q \Rightarrow P$ is a tautology, but $P \wedge (\neg P)$ is a contradiction. The negation of any tautology is a contradiction.

Definition 1.8. *Two statements S_1 and S_2 are logically equivalent exactly when $S_1 \Leftrightarrow S_2$ is a tautology. We will write $S_1 \equiv S_2$ if S_1 and S_2 are logically equivalent.*

For example, $P \wedge P \equiv P$ and $P \wedge Q \equiv Q \wedge P$. We summarize the basic logical equivalences in the next theorem. In particular, we learn how to negate a conjunction or disjunction of statements using the so-called De Morgan's laws.

Theorem 1.9. *We have*

    a) Associative laws: $P \wedge (Q \wedge R) \equiv (P \wedge Q) \wedge R$, $P \vee (Q \vee R) \equiv (P \vee Q) \vee R$.

    b) Commutative laws: $P \Leftrightarrow Q \equiv Q \Leftrightarrow P$, $P \wedge Q \equiv Q \wedge P$, $P \vee Q \equiv Q \vee P$.

    c) Idempotency laws: $P \wedge P \equiv P$, $P \vee P \equiv P$.

    d) Absorption laws: $P \wedge (P \vee Q) \equiv P$, $P \vee (P \wedge Q) \equiv P$.

    e) Distributive laws: $P \wedge (Q \vee R) \equiv (P \wedge Q) \vee (P \wedge R)$, $P \vee (Q \wedge R) \equiv (P \vee Q) \wedge (P \vee R)$.

    f) Law of double negation: $\neg(\neg P) \equiv P$.

    g) De Morgan's laws: $\neg(P \wedge Q) \equiv (\neg P) \vee (\neg Q)$, $\neg(P \vee Q) \equiv (\neg P) \wedge (\neg Q)$.

    h) Contrapositive law: $P \Rightarrow Q \equiv (\neg Q) \Rightarrow (\neg P)$.

Proof. The method of proof of this theorem is to check that the truth tables of various statements are the same. We will illustrate this with part g; the other parts are similar.

P	Q	$P \wedge Q$	$\neg(P \wedge Q)$	$\neg P$	$\neg Q$	$(\neg P) \vee (\neg Q)$
T	T	T	F	F	F	F
T	F	F	T	F	T	T
F	T	F	T	T	F	T
F	F	F	T	T	T	T

P	Q	$P \vee Q$	$\neg(P \vee Q)$	$\neg P$	$\neg Q$	$(\neg P) \wedge (\neg Q)$
T	T	T	F	F	F	F
T	F	T	F	F	T	F
F	T	T	F	T	F	F
F	F	F	T	T	T	T

$\square$

When a long sentence contains one or more parts that are themselves compound sentences, parentheses may be needed. For example, $(\neg P) \wedge (P \vee Q)$ is different from $\neg (P \wedge (P \vee Q))$. You can check this by looking at their truth tables. As a general rule, the conjunction takes precedence over the disjunction: when we write $P \wedge Q \vee R$ we mean $(P \wedge Q) \vee R$. This is similar to multiplication being done before addition.

When we write proofs, it is important to make valid arguments. We say that the statement B is a valid consequence of the statements $A_1, A_2, ..., A_n$ if for every assignment of truth values that makes all the statements $A_1, A_2, ..., A_n$ true, the statement B is also true.

For example, if A_1 is the statement "x is odd", A_2 is "y is odd", and B is "$x + y$ is even", then $A_1 \wedge A_2 \Rightarrow B$. Be careful; if we start with wrong premises or we use wrong reasoning, we may end with wrong conclusions.

Example 1.10. *Consider the statement "If $x = 1$ then $x = 0$" with the following "proof": Multiplying both sides of the equation $x = 1$ by x we obtain $x^2 = x$, hence $x^2 - x = 0$. Factoring, we get $x(x - 1) = 0$. Dividing by $x - 1$ yields the desired conclusion $x = 0$. The flaw in the argument comes from the fact that for $x = 1$, $x - 1$ becomes 0 and we cannot divide by 0.*

Example 1.11. *Negate the following statements:*
a) If I go to the party tonight, then she is there.
b) The number n is even and n is divisible by 4.
c) A function f is continuous or it is differentiable.
d) If n is an integer, then n is divisible by 3.

Solution. a) The statement is a conditional of the form $P \Rightarrow Q$, where P is "I go to the party tonight" and Q is "she is there". We know that $\neg (P \Rightarrow Q) \equiv (P \wedge \neg Q)$. Hence the negation is: "I go to the party tonight and she is not there".

b) Here we use one of the De Morgan laws: $\neg (P \wedge Q) \equiv (\neg P) \vee (\neg Q)$. The negation is: The number n is not even (i.e., is odd) or is not divisible by 4.

c) The negation of a disjunction goes like $\neg (P \vee Q) \equiv (\neg P) \wedge (\neg Q)$. We obtain: A function f is not continuous and it is not differentiable.

d) Again we negate a conditional. We obtain: n is an integer and n is not divisible by 3.

1.4 Quantifiers

Many mathematical statements use quantifiers. As we mentioned before, the symbol $\forall$ is the universal quantifier meaning "for all" or "for every", and $\exists$ is the existential quantifier meaning "there is" or "there exist". Both quantifiers refer to a certain "universe" of the discourse, a set of numbers or symbols which must be specified or be clear from the context. We will talk more about the concept of universe in the chapter about sets. The symbol $\exists! x$ means "there exists a unique x". A statement like "For every positive real number x there is a real number y such that $y^2 = x$" becomes

$$\forall x \in (0, \infty) \, \exists \, y \in \mathbb{R} : y^2 = x.$$

The universe for x is the set of positive real numbers, and for y the set of real numbers. This is a true statement since one can take $y = \sqrt{x}$ or $y = -\sqrt{x}$. On the other hand,

$$\forall x \in (0, \infty) \, \exists! \, y \in \mathbb{R} : y^2 = x$$

is false, since y is not unique: for $x = 1$ we can take $y = -1$ or $y = 1$. The order of quantifiers is very important, since for example

$$\exists y \in \mathbb{R} \, \forall x \in (0, \infty) : y^2 = x$$

is a false statement: once you fix a real number y, the number y^2 cannot be equal with all positive numbers.

We already learned how to negate conjunctions, disjunctions and conditionals. How do you negate a statement involving quantifiers? We will see that $\forall$ becomes $\exists$ and $\exists$ becomes $\forall$.

Example 1.12. *To negate "All horses are black", we can say "Not all horses are black" or "There are horses which are not black".*

Example 1.13. *Suppose the universe is the set of rational numbers and let's negate the statement "For every rational number x there exists an integer n that is greater than x", in symbols*

$$\forall x \, \exists n : n > x.$$

Since not for every rational number x we can find an integer n such that $n > x$, we must be able to find some rational number x such that for all integers n we have $n \leq x$, in symbols,

$$\exists x \, \forall n : n \leq x.$$

In general, we have the following basic negation rules for quantifiers:

Theorem 1.14. *(De Morgan's laws for quantifiers)*

$$\neg(\forall x \, P(x)) \equiv \exists x \, \neg P(x), \quad \neg(\exists x \, P(x)) \equiv \forall x \, \neg P(x).$$

Example 1.15. *The negation of the true statement*

$$\forall x \in (0, \infty) \, \exists \, y \in \mathbb{R} : y^2 = x$$

is the false statement "There exists a positive number x such that for all real y we have $y^2 \neq x$", in symbols,

$$\exists x \in (0, \infty) \, \forall y \in \mathbb{R} : y^2 \neq x.$$

Example 1.16. *Assume x_n for $n \geq 1$ and x are real numbers. The negation of*

$$\forall \varepsilon > 0 \; \exists N \geq 1 \; \forall n \geq N \text{ we have } |x_n - x| < \varepsilon$$

is

$$\exists \varepsilon > 0 \; \forall N \geq 1 \; \exists n \geq N \text{ such that } |x_n - x| \geq \varepsilon.$$

Definition 1.17. *Suppose that the statement $\forall x(P(x) \Rightarrow Q(x))$ is false, therefore the statement $\exists x \; \neg(P(x) \Rightarrow Q(x))$ is true. A value a such that $\neg(P(a) \Rightarrow Q(a))$ is true will be called a* counterexample *for $\forall x \; (P(x) \Rightarrow Q(x))$.*

Example 1.18. *The statement "For all integers n, if n is a multiple of 4, then n is a multiple of 12" is false, since there exists $n = 8$, which is a multiple of 4, but not a multiple of 12. The number $n = 8$ is a counterexample to our statement.*

Note that the statement $\exists! x \; P(x)$ is equivalent to "There is an x such that $P(x)$ is true, and for all y such that $P(y)$ is true, it follows that $y = x$", in symbols,

$$(\exists x \; P(x)) \wedge (\forall y \; (P(y) \Rightarrow y = x)).$$

Therefore, the negation of $\exists! x \; P(x)$ is "For all x we have that $P(x)$ is false or there is y for which $P(y)$ is true and $y \neq x$", in symbols,

$$(\forall x(\neg P(x))) \vee (\exists y(P(y) \wedge y \neq x)).$$

Working with several quantifiers requires a lot of care.

Example 1.19. *Consider the statement $P(x, y) : x+y = 1$. Then $\forall x \exists y P(x, y)$ is true, because given an x we can choose $y = 1 - x$, but $\exists y \forall x P(x, y)$ is false, since there is no y good for all x such that $x + y = 1$.*

Example 1.20. *Compare the following statements for a function $f : \mathbb{R} \to \mathbb{R}$.*

1) For any positive ε and any real number x there is a positive δ such that for any $h \in (-\delta, \delta)$ we have $f(x + h) \in (f(x) - \varepsilon, f(x) + \varepsilon)$, in symbols

$$\forall \varepsilon > 0 \; \forall x \in \mathbb{R} \; \exists \delta > 0 \; \forall h \in (-\delta, \delta) \Rightarrow f(x + h) \in (f(x) - \varepsilon, f(x) + \varepsilon).$$

2) For any positive ε there is a positive δ such that for any real number x and for any $h \in (-\delta, \delta)$ we have $f(x + h) \in (f(x) - \varepsilon, f(x) + \varepsilon)$, in symbols,

$$\forall \varepsilon > 0 \; \exists \delta > 0 \; \forall x \in \mathbb{R} \; \forall h \in (-\delta, \delta) \Rightarrow f(x + h) \in (f(x) - \varepsilon, f(x) + \varepsilon).$$

They may look similar, but the order of quantifiers makes a big difference: the point is that in statement 1, the number δ depends both on x and ε, but in statement 2, the number δ depends only on ε, so it is good for all values of x. In analysis, statement 1 defines the notion of pointwise continuity of a function, and statement 2 defines the notion of uniform continuity.

1.5   Exercises

Exercise 1.21. *You may have heard the following puzzle about the farmer who needs to cross a river with a wolf, a goat, and a cabbage. There is a boat that can fit himself plus either the wolf, the goat, or the cabbage. But if the wolf and the goat are left alone, the wolf will eat the goat. Also, if the goat and the cabbage are alone, the goat will eat the cabbage. How can this person take the wolf, the goat, and the cabbage across the river?*

Exercise 1.22. *A scientist gathered four math students. They were then lined up so that each one could see the one in front of them but not behind them. Each had a hat placed on their head.*

So the student in the back could see the hats of the three students in front, but the student in front could not see any hats. "There is a red hat, a white hat, a blue hat, and a hat that is a duplicate of one of those colors", the scientist said.

Starting with the one in the back, each student was asked what color hat they were wearing. They all gave the correct answer!

What was the arrangement of the hats that made this possible?

Exercise 1.23. *In the following conditionals, identify the antecedent and the consequent:*
  a) If it is snowing, I will go to work.
  b) The number n is odd only if it is prime.
  c) I wake up if the baby cries.
  d) Whenever the alarm goes off, I leave the room.

Exercise 1.24. *Consider this sentence:*

$$\text{If } 2 < 3 \text{ then } 1 + 1 = 2 \text{ or } 3 + 2 = 6 \text{ and } 5 > 7.$$

This is impossible to read accurately in this form. Use parentheses to construct four meaningful statements, and determine if they are true or false.

Exercise 1.25. *Assume x to be a real number. The statement "If $x < 0$, then $x^2 > 0$" is certainly correct. Now substitute $x = -1$, $x = 0$, and $x = 1$ to obtain three conditionals, and explain why they are true.*

Exercise 1.26. *Find truth tables for each of the following:*
  a) $\neg(P \wedge Q)$.
  b) $\neg[(P \vee Q) \wedge ((\neg P) \vee (\neg Q))]$.
  c) $\neg(P \vee Q) \vee \neg(Q \wedge P)$.
  d) $P \Rightarrow (Q \Rightarrow P)$.
  e) $(P \Rightarrow Q) \Rightarrow ((\neg Q) \Rightarrow (\neg P))$.
  f) $(P \Rightarrow Q) \Leftrightarrow ((\neg P) \vee Q)$.

Exercise 1.27. *Show that the following are tautologies:* $P \vee \neg P$, $P \Rightarrow P$, $P \Leftrightarrow P$.

Exercise 1.28. *Which of the following statements are tautologies?*
  a) $((P \Rightarrow Q) \wedge P) \Rightarrow Q$.
  b) $(P \Rightarrow Q) \vee (Q \Rightarrow P)$.
  c) $((P \Rightarrow Q) \Rightarrow Q) \Rightarrow P$.
  d) $(P \Rightarrow Q) \Rightarrow ((\neg P) \Rightarrow (\neg Q))$.

Exercise 1.29. *Consider the statement "If n is odd, then $n^2 - n - 6$ is even". State the contrapositive and the converse statements.*

Exercise 1.30. *Construct a truth table for* $(P \vee Q) \Rightarrow (P \wedge \neg Q)$. *Find a simpler proposition that is logically equivalent with* $(P \vee Q) \Rightarrow (P \wedge \neg Q)$.

Exercise 1.31. *Are the following arguments valid?*
  a) If a function f is differentiable, then f is continuous. Assume f is continuous. Therefore, f is differentiable.
  b) If f is not continuous, then f is not differentiable. Assume f is differentiable. Therefore, f is continuous.
  c) If a function f is differentiable, then f is continuous. Assume f is not differentiable. Therefore f is not continuous.

Exercise 1.32. *Suppose we are given the following facts:*
  (a) I will be admitted to the university only if I am smart.
  (b) If I am smart, then I do not have to work hard.
  (c) I have to work hard.
  What can be deduced?

Exercise 1.33. *Each of the following statements is a result from calculus. In each case, write the contrapositive of the statement and the converse of the statement. Is the converse of the statement true?*
  1) If the series $\sum_{n=1}^{\infty} a_n$ is convergent, then $\lim_{n \to \infty} a_n = 0$.
  2) If f is constant on $[a, b]$, then $f'(x) = 0$ on $[a, b]$.
  3) If the series $\sum_{n=1}^{\infty} |a_n|$ converges, then $\sum_{n=1}^{\infty} a_n$ converges.
  4) If f is continuous on $[a, b]$, then f attains a maximum value on $[a, b]$.

Exercise 1.34. *Write the following statements and their negations using variables and quantifiers. Then decide which statements are true.*
  a) Any integer is odd.
  b) There is a rational number r such that $r^2 = 2$.
  c) For any real number x, there is a unique real number y such that $y^3 = x$.
  d) For any odd integer n, there is an even integer m such that m divides n.
  e) If x is a rational number, then $x^2 + 1 \neq 0$.

f) There exists an x such that for some y the equality $x = 3y$ holds.

g) If x and y are any real numbers and for all $\varepsilon > 0$ we have $x < y + \varepsilon$, then $x \leq y$.

h) The square root of any irrational number is irrational.

2

Proofs: Structures and strategies

When you study physics, economics, psychology, mathematics, or any other subject, there are certain key words which you must learn to use correctly. You may not completely understand them when you begin, but continued usage helps you become familiar with them. In some cases, a perfectly precise definition of a term may never be available, but with experience you may be able to understand and use the idea represented by the term. For example, it is very difficult to pin down exactly what "force" is, but enough is known about force that all sorts of things can be proved about it.

In the same way, when we investigate the structure of a mathematical theory, there are always some concepts which are so basic that they do not admit any simple definition. They can really only be understood by the properties they have or, putting it another way, by statements which use them correctly. These basic concepts are referred to by words or symbols that we call "primitive" or "undefined" terms.

The most basic statements we write down (using the undefined terms) are the *postulates* or *axioms*. These are often understood to be so obvious that no proof is required, though in a sense that is inadequate as a description. More specifically, we will here understand postulates as statements which are not necessarily true or false, but which are the basis for an ongoing reasoning process.

In a sense, while we are developing a theory, we regard the postulates as true, but whether they really are or not is generally irrelevant to the theory. The typical daily project in mathematics is to discover a consequence of one or more hypotheses and, hopefully, to prove it. In practice, that's what mathematics is all about.

In this chapter, we discuss different methods of proofs, illustrating the rules of logic that we have learned in the previous chapter. We begin with the concepts of axiom, theorem, proof, and conjecture. We will discuss several examples of direct proofs, indirect proofs, and proofs by contradiction. We also illustrate proofs by cases, existence proofs, proofs by counterexample, and proofs by mathematical induction. More examples of proofs will appear in the subsequent chapters, after new concepts are introduced.

2.1 Axioms, theorems and proofs

In constructing a mathematical theory, we often start with some statements called *axioms* or *postulates* which are accepted to be true, and using valid rules of logic we prove new true statements, the *theorems*. In the process, we may have to prove auxiliary results, called *lemmas*. A direct consequence of a theorem is called a *corollary*.

You already have applied the Pythagorean theorem in geometry or the fundamental theorem of calculus many times. A proof is a deductive argument and it may be based on previously established statements. A proof could be straightforward, by just doing a computation. For example, $\int_0^\pi \sin x dx = 2$ is a true statement, and becomes a theorem after we compute the integral using the fundamental theorem of calculus,

$$\int_0^\pi \sin x \, dx = -\cos x \Big|_0^\pi = -\cos(\pi) + \cos(0) = -(-1) + 1 = 2.$$

The statement "A triangle with two congruent angles is isosceles" is a proposition which becomes a theorem after we prove that two sides of the triangle are congruent.

Sometimes a proof can be long and complicated, requiring smart ideas and tricks, connecting several areas of mathematics. One way or another, a proof should be based on valid arguments. You will encounter in this chapter many methods and examples of proofs. Read them carefully and try to understand each step. You will need to practice proofs by solving the exercises.

The symbol $\square$ indicates the end of a proof. Some people prefer to mark the end of a proof with the letters Q.E.D. (*quod erat demonstrandum*).

A *conjecture* is a statement believed to be true, for which no proof has been provided.

Recall that a divides b (or b is divisible by a or a is a divisor of b) if $b = au$ for some integer u. We write $a \mid b$ when a divides b.

The Goldbach conjecture states that every even integer larger or equal to 4 is the sum of two prime numbers (a positive integer ≥ 2 is prime if the only divisors are 1 and the number itself; the number 1 is not a prime). Here are a couple of primes:

$$2, 3, 5, 7, 11, 13, 17, 19, 23, 29, 31, 37, 41, 43, 47, 53, \ldots$$

We have

$$4 = 2+2, \ 6 = 3+3, \ 8 = 5+3, \ 10 = 3+7 = 5+5, 12 = 5+7, 14 = 3+11 = 7+7, \ldots$$

Nobody was able to prove the Goldbach conjecture yet. A conjecture is sometimes called a hypothesis, like the famous unsolved Riemann hypothesis.

The conjectures may be settled by finding a proof, or by finding a counterexample, in which case the conjecture is false. The statement "For each nonnegative integer n, the Fermat number $F_n = 2^{2^n} + 1$ is prime" used to be a conjecture. Indeed, it works for $n = 0, 1, 2, 3, 4$ because $F_0 = 3, F_1 = 5, F_2 = 17, F_3 = 257, F_4 = 65537$ are prime numbers (we will see later how to verify this). But Euler showed in 1732 that $F_5 = 4294967297$ is divisible by 641, hence F_5 is not a prime, therefore the conjecture is false. We say that $n = 5$ provides a counterexample to the conjecture.

Example 2.1. *As an abstract example of a theory, suppose that we construct statements which are words using only the symbols a, b, S. Suppose S is the only axiom, and the rules are that we can replace S by aSb, and that we can delete an S. Is $aabb$ a true statement? How about SSa?*

It can actually be proved that the only theorems are $\underbrace{a...a}_{n} S \underbrace{b...b}_{n}$ and $\underbrace{a...a}_{n} \underbrace{b...b}_{n}$ for $n \geq 0$. Therefore, $aabb$ is a true statement, and SSa is false.

We will give later examples of axioms used to define positive integers and to prove their properties. Also, we will give axioms for set theory.

An open sentence (or *predicate*) is a sentence depending on one or more variables, which becomes true or false when we fix the variable.

For example, consider the sentence $P(a)$: the real function $f(x) = |x|$ is differentiable at a. Then $P(1)$ is a true proposition, since $f'(1) = 1$ but $P(0)$ is false since $f'(0)$ does not exist. Can you explain why?

Recall that $|x|$ denotes the *absolute value* of x and

$$|x| = \begin{cases} x, & x \geq 0 \\ -x, & x < 0. \end{cases}$$

We will use the universal quantifier $\forall$ and the existential quantifier $\exists$ discussed in the previous chapter to form new statements, using open sentences like

$$\forall x \, P(x), \quad \exists y \, Q(y), \quad \forall x \, \exists y \, R(x, y).$$

Recall that the expression $\exists! \, x \, P(x)$ means that there is a unique x such that $P(x)$.

2.2 Direct proof

The pattern of a *direct proof* is as follows: suppose P is true and $P \Rightarrow Q$ is true. Then Q is true. This is also called the rule of *modus ponens*. If we deal with theorems in the form of conditional statements, a direct proof might be appropriate.

Example 2.2. *If two integers are odd, then their product is odd.*

Proof. Since a, b are odd, we have $a = 2m+1, b = 2n+1$ for some integers m, n. Then $ab = (2m+1)(2n+1) = 4mn+2m+2n+1 = 2(2mn+m+n)+1 = 2k+1$, where $k = 2mn + m + n$. Hence ab is odd. $\square$

Exercise 2.3. *Use the above example to prove: If x is odd, then x^2 is odd.*

Example 2.4. *If x, y are positive real numbers, then $x + y \geq 2\sqrt{xy}$.*

Proof. We rewrite the inequality as $x + y - 2\sqrt{xy} \geq 0$. This inequality is equivalent to $(\sqrt{x})^2 + (\sqrt{y})^2 - 2\sqrt{x}\sqrt{y} \geq 0$ or $(\sqrt{x} - \sqrt{y})^2 \geq 0$, which is true. $\square$

Notice that we used the fact that x, y are positive. For $x = y = -1$, check that the inequality is false.

Example 2.5. *Let $a \neq 0$. If $b^2 - 4ac \geq 0$, prove that the quadratic equation $ax^2 + bx + c = 0$ has real roots given by the familiar formula*

$$x = \frac{-b \pm \sqrt{b^2 - 4ac}}{2a}.$$

Proof. Dividing the equation by a we get $x^2 + \dfrac{b}{a}x + \dfrac{c}{a} = 0$. The idea is to complete the square, by adding and subtracting what is needed:

$$x^2 + 2\frac{b}{2a}x + \left(\frac{b}{2a}\right)^2 + \frac{c}{a} - \left(\frac{b}{2a}\right)^2 = 0,$$

$$\left(x + \frac{b}{2a}\right)^2 = \frac{b^2 - 4ac}{4a^2}.$$

Since $b^2 - 4ac \geq 0$, we can solve the last equation and get

$$x + \frac{b}{2a} = \pm\frac{\sqrt{b^2 - 4ac}}{2a}, \quad x = -\frac{b}{2a} \pm \frac{\sqrt{b^2 - 4ac}}{2a}.$$

$\square$

We illustrate now with a direct proof in geometry.

Theorem 2.6. *Let ABC be a right triangle with hypotenuse BC. Let D be on BC such that $AD \perp BC$. Then $AB^2 = BD \cdot BC$ and $AC^2 = DC \cdot BC$.*

Proof.

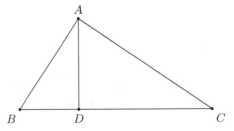

The triangles ADB and BAC are similar, since they have congruent angles. Also, the triangles ADC and BAC are similar. It follows that

$$\frac{AB}{BD} = \frac{BC}{AB} \quad \text{and} \quad \frac{AC}{DC} = \frac{BC}{AC},$$

hence $AB^2 = BD \cdot BC$ and $AC^2 = DC \cdot BC$. □

Corollary 2.7. *(Pythagorean theorem). Let ABC be a right triangle with hypotenuse BC. Then $AB^2 + AC^2 = BC^2$.*

Proof. We have

$$AB^2 + AC^2 = BD \cdot BC + DC \cdot BC = (BD + DC) \cdot BC = BC \cdot BC = BC^2.$$

□

2.3 Contrapositive proof

A conditional statement of the form $P \Rightarrow Q$ may also be proved indirectly, using a *contrapositive proof*. The idea is to prove the logically equivalent statement $(\neg Q) \Rightarrow (\neg P)$, which sometimes is easier.

Example 2.8. *Let n be an integer. If n^2 is odd, then n is odd.*

Proof. A direct proof would be awkward, since from $n^2 = 2k + 1$ we don't know where to go. We prove instead: if n is even, then n^2 is even, which is easier. Indeed, $n = 2k$ implies $n^2 = 4k^2 = 2 \cdot 2k^2$, which is even. □

Example 2.9. *Let x, y be integers. If 3 does not divide xy, then 3 does not divide x and 3 does not divide y.*

Proof. By De Morgan's laws of negation (see Theorem 1.9 part g), we prove: if $3 \mid x$ or $3 \mid y$, then $3 \mid xy$. In the case $x = 3a$, we get $xy = 3ay$ and $3 \mid xy$. In the second case, $y = 3b$ and $xy = 3xb$, so $3 \mid xy$. □

Exercise 2.10. *Use the method of contrapositive proof to prove the following. In each case, think also of a direct proof, if possible, and compare them:*

a. Let x be an integer. If $x^2 + 4x < 0$, then $x < 0$.

b. If both ab and $a + b$ are odd integers, then a and b have different parity (one is odd and one is even).

c. Let a be an integer. If a^2 is not divisible by 4, then a is odd.

2.4   Proof by contradiction

Another indirect method of proving conditional statements $P \Rightarrow Q$ is the *proof by contradiction*. We assume that $P \Rightarrow Q$ is false, and try to get a contradiction, usually a statement like $R \wedge (\neg R)$. Recall that $P \Rightarrow Q$ is logically equivalent to $(\neg P) \vee Q$, so its negation is $P \wedge (\neg Q)$.

Example 2.11. *If a, b are integers, then $a^2 - 4b \neq 2$.*

Proof. Assume the implication is false, namely that there exist integers a, b such that $a^2 - 4b = 2$. We get $a^2 = 4b + 2 = 2(2b + 1)$, so a must be even, say $a = 2c$. Plugging $a^2 - 4b = 2$ back in the equality we get $4c^2 - 4b = 2$ or $4(c^2 - b) = 2$, which says that 2 is a multiple of 4, a contradiction. Something went wrong, so it must be that for all integers a, b we have $a^2 - 4b \neq 2$. $\square$

In fact, the proof by contradiction can be applied to other statements, not necessarily conditional statements.

Example 2.12. *Let's prove that the number $\sqrt{2}$ is irrational (not of the form a/b with a, b integers and $b \neq 0$).*

Proof. First recall that $\sqrt{2}$ is a positive number such that $(\sqrt{2})^2 = 2$. Assume $\sqrt{2}$ is rational, hence $\sqrt{2} = a/b$, with a, b integers and $b \neq 0$. By simplifying the fraction, we may assume that a and b are relatively prime (their greatest common divisor is 1). We get $a = b\sqrt{2}$. Squaring both sides, $a^2 = 2b^2$, which implies that a is even, say $a = 2n$. But then $4n^2 = 2b^2$, hence $b^2 = 2n^2$ and b must also be even, a contradiction with the assumption that a and b are relatively prime. $\square$

Example 2.13. *There are infinitely many prime numbers (recall that an integer $p \geq 2$ is prime if the only divisors are 1 and p).*

Proof. Looking for a contradiction, suppose there are only finitely many primes, call them $p_1, p_2, ..., p_n$, where $p_1 < p_2 < ... < p_n$, so p_n is the largest. Consider the number $a = p_1 p_2 \cdots p_n + 1$. Like any natural number, a has a prime divisor, say p_k. We get

$$p_1 p_2 \cdots p_k \cdots p_n + 1 = c p_k.$$

Dividing both sides by p_k, it follows that

$$\frac{p_1 p_2 \cdots p_n}{p_k} + \frac{1}{p_k} = c,$$

which implies that $1/p_k$ is an integer, a contradiction of the fact that $p_k \geq 2$. $\square$

Example 2.14. *Let ABC be an equilateral triangle. Let a, b be two lines in the plane such that a passes through A and a ⊥ AC, b passes through C and b ∥ AB. Prove that the lines a, b intersect in a point.*

Proof. If a, b don't intersect, then $a \parallel b$, and since $b \parallel AB$, it follows that $a = AB$ since a contains A. But then $\widehat{BAC}$ is a right angle since $a \perp AC$, a contradiction with the fact that ABC is equilateral. □

Exercise 2.15. *Prove by contradiction the following statements. In each case, think also about a direct proof or a contrapositive proof, if possible.*
  a. If a, b are integers, then $a^2 - 4b \neq 3$.
  b. The number $\sqrt{6}$ is irrational.
  c. The number $\sqrt{2} + \sqrt{3}$ is irrational.
  d. If a is rational and ab is irrational, then b is irrational.
  e. If $a, b, \sqrt{a} + \sqrt{b}$ are rational numbers, then $\sqrt{a}, \sqrt{b}$ are also rational.

2.5 Proofs of equivalent statements

How about the proof of a biconditional statement $P \Leftrightarrow Q$? There are two steps: we prove first $P \Rightarrow Q$ and then $Q \Rightarrow P$. For these, we may use any method we learned so far: direct proof, contrapositive proof or proof by contradiction.

Example 2.16. *Suppose a, b are integers. Then $10 \mid a - b$ if and only if $2 \mid a - b$ and $5 \mid a - b$.*

Proof. We use a direct proof for both implications. If $10 \mid a - b$, then certainly $2 \mid a - b$ and $5 \mid a - b$, since 2 and 5 divide 10.

Assuming that $2 \mid a - b$ and $5 \mid a - b$, we get $a - b = 2c = 5d$ for some c, d. Since 2 and 5 are relatively prime, it must be that 5 divides c, hence $c = 5e$ and $a - b = 10e$. □

Sometimes, we may have to prove that several statements are equivalent, like

$$P \Leftrightarrow Q \Leftrightarrow R \Leftrightarrow S.$$

We can use a circle of implications like $P \Rightarrow Q \Rightarrow R \Rightarrow S \Rightarrow P$ or split the equivalences into smaller groups.

Example 2.17. *Consider a triangle ABC. The following are equivalent*
  1) The triangle is isosceles with $AB = AC$.
  2) The angles $\widehat{ABC}$ and $\widehat{ACB}$ are congruent.
  3) The altitude AD is a median, i.e., $BD = DC$.
  4) The altitude AD is a bisector of $\widehat{BAC}$.
  5) The altitudes BE and CF are congruent.
  6) The medians BN and CM are congruent.

Proof.

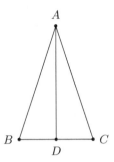

1 ⇒ 2. Assuming $AB = AC$, consider the altitude $AD \perp BC$. The triangles ADB and ADC are right triangles with a common leg and congruent hypotenuses, so they are congruent. In particular, $\widehat{ABD} = \widehat{ACD}$, so $\widehat{ABC} = \widehat{ACB}$.

2 ⇒ 3. Consider $AD \perp BC$. Since $\widehat{ABC} = \widehat{ACB}$, the right triangles ADB and ADC have congruent angles and a common leg, so they are congruent. In particular $BD = DC$.

3 ⇒ 4. Since the altitude AD is a median, we have $BD = DC$. The right triangles ADB and ADC have congruent legs, hence they are congruent. In particular $\widehat{BAD} = \widehat{CAD}$ and AD is a bisector of $\widehat{BAC}$.

4 ⇒ 1. Since the altitude AD is a bisector, the right triangles ADB and ADC have congruent angles and a common leg, so they are congruent. In particular $AB = AC$.

We just proved that the first four conditions are equivalent.

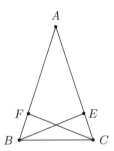

2 ⇒ 5. Consider the altitudes $BE \perp AC$ and $CF \perp AB$. The right triangles BEC and CFB have a common hypotenuse and $\widehat{BCE} = \widehat{CBF}$, so they are congruent. It follows that $BE = CF$.

5 ⇒ 2. Since the altitudes BE and CF are congruent, the right triangles BEC and CFB have a common hypotenuse and congruent legs, so they are congruent. In particular $\widehat{BCE} = \widehat{CBF}$ and $\widehat{ABC} = \widehat{ACB}$.

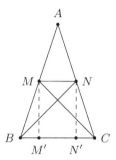

1 ⇒ 6. Consider the midpoints M of AB and N of AC. Since $AB = AC$, we get $BM = CN$. Since we also have $\widehat{MBC} = \widehat{NCB}$, it follows that the triangles MBC and NCB are congruents, in particular $BN = CM$.

6 ⇒ 1. Assume that the medians BN and CM are congruent. Since M and N are midpoints of AB and AC respectively, it follows that $MN \parallel BC$. Let $MM' \perp BC$ and $NN' \perp BC$. It follows that $MM'N'N$ is a rectangle, so $MM' = NN'$. Since $BN = CM$, the right triangles $BN'N$ and $CM'M$ are congruent, so $\widehat{NBN'} = \widehat{MCM'}$. It follows that the triangles MBC and NCB are congruent (case side-angle-side) and therefore $BM = CN$, so $AB = AC$. □

2.6   Proof by cases

Some proofs are based on analyzing all possible cases.

Example 2.18. *Prove that for all real numbers x, y we have $|xy| = |x||y|$.*

Proof. Recall that
$$|x| = \begin{cases} x, & x \geq 0 \\ -x, & x < 0. \end{cases}$$
There are four cases to consider:

1) $x \geq 0$ and $y \geq 0$. In this case $xy \geq 0$, so $|xy| = xy = |x||y|$.

2) $x \geq 0$ and $y < 0$. In this case $xy \leq 0$, so $|xy| = -(xy) = x(-y) = |x||y|$.

3) $x < 0$ and $y < 0$. In this case $xy > 0$, so $|xy| = xy = (-x)(-y) = |x||y|$.

4) $x < 0$ and $y \geq 0$. In this case $xy \leq 0$, so $|xy| = -(xy) = (-x)y = |x||y|$. □

Example 2.19. *Prove that the set of real solutions of $|x - 1| < |x - 3|$ is the interval $(-\infty, 2)$.*

Proof. We have

$$|x-1| = \begin{cases} x-1 & \text{if} \quad x \geq 1 \\ -x+1 & \text{if} \quad x < 1. \end{cases} \quad , \quad |x-3| = \begin{cases} x-3 & \text{if} \quad x \geq 3 \\ -x+3 & \text{if} \quad x < 3. \end{cases}$$

Case $x \geq 3$. The inequality becomes $x - 1 < x - 3$, so $-1 < -3$, which is false. This case gives no solution.

Case $1 \leq x < 3$. The inequality becomes $x - 1 < 3 - x$, so $2x < 4$, or $x < 2$. We get $[1, 2)$ as a solution set.

Case $x < 1$. We obtain $1 - x < 3 - x$, or $1 < 3$, which is true. We also get $(-\infty, 1)$ as part of the solution.

The conclusion is that $x \in [1, 2)$ or $x \in (-\infty, 1)$, so $x \in (-\infty, 2)$. $\square$

Example 2.20. *Show that n^4 ends in $0, 1, 5$ or 6 for any positive integer n.*

Proof. Indeed, let's first find the last digit of n^4 for $n \in \{0, 1, 2, ..., 9\}$. We have $0^4 = 0, 1^4 = 1, 2^4 = 16, 3^4 = 81, 4^4 = 256, 5^4 = 625, 6^4 = 36 \cdot 36$ ends in 6, $7^4 = 49 \cdot 49$ ends in 1, $8^4 = 64 \cdot 64$ ends in 6, $9^4 = 81 \cdot 81$ ends in 1. We conclude that the last digit of n^4 is $0, 1, 5$ or 6 for any positive integer n. $\square$

Exercise 2.21. *Prove the triangle inequality: for all x, y in $\mathbb{R}$ we have*

$$|x + y| \leq |x| + |y|.$$

2.7   Existence proofs

To prove a statement of the form $\exists x P(x)$, where x is a number, one can try to construct directly a value x_0 such that the statement $P(x_0)$ is valid.

Example 2.22. *Show that there is a real number x such that $x^2 = 3$.*

Proof. We can directly check that $x_0 = \sqrt{3}$ works. Of course, $x_1 = -\sqrt{3}$ is another choice. $\square$

Exercise 2.23. *Prove that there is a prime number between 100 and 200.*

Another method to prove $\exists x P(x)$ is by contradiction: we assume that the negation of the statement holds; in other words, assume that $\forall x \neg P(x)$, and derive a contradiction.

Recall that the intermediate value theorem from calculus says that a continuous function $f : [a, b] \to \mathbb{R}$ takes all the values between $f(a)$ and $f(b)$. In particular, if a continuous function has values of opposite signs at the endpoints of an interval, then it has a root in that interval.

This theorem could be used to give existence proofs which are nonconstructive in the sense that we do not specify the value of x, like in the following example.

Example 2.24. *Prove that the equation $x^3 = 12x + 5$ has a root in the interval $[-2, 2]$.*

Proof. Suppose that for any $x \in [-2, 2]$ we have $x^3 \neq 12x + 5$. Consider the continuous function $f : [-2, 2] \to \mathbb{R}$, $f(x) = x^3 - 12x - 5$. The assumption implies that $f(x) \neq 0$ for all $x \in [-2, 2]$. But $f(-2) = 11 > 0$ while $f(2) = -21 < 0$, which contradicts the intermediate value theorem on the interval $[-2, 2]$. □

Example 2.25. *Prove that there is a positive integer n such that $2^{2^n} + 1$ is not a prime.*

Proof. It turns out that

$$2^{2^1} + 1 = 5, 2^{2^2} + 1 = 17, 2^{2^3} + 1 = 257, 2^{2^4} + 1 = 65537$$

are primes. Indeed, it is easy to see that 5 and 17 are primes. To check that a bigger number n is prime, we will prove in Theorem 6.38 that it suffices to check that n is not divisible by any prime p with $2 \leq p \leq \sqrt{n}$. Since 257 is not divisible by $2, 3, 5, 7, 11$ and 13, it follows that 257 is prime. Note that $16 < \sqrt{257} < 17$ since $16^2 = 256$ and $17^2 = 289$. For 65537 we need to check that it is not divisible by all primes less or equal to 256, since $256 < \sqrt{65537} < 257$. These primes are

$$2, 3, 5, 7, 11, 13, 17, 19, 23, 29, 31, 37, 41, 43, 47, 53, 59, 61, 67, 71, 73, 79,$$

$$83, 89, 97, 101, 103, 107, 109, 113127, 131, 137, 139, 149, 151, 157, 163, 167, 173,$$

$$179, 181, 191, 193, 197, 199, 211, 223, 227, 229, 233, 239, 241, 251.$$

The tedious task can be done by using a calculator. Now it can be checked that

$$2^{2^5} + 1 = 4294967297 = 641 \cdot 6700417$$

is divisible by 641, so the statement is true for $n = 5$. The numbers $F_n = 2^{2^n} + 1$ are the famous Fermat numbers. □

Although an example is sufficient to prove an existence statement, this is not the case for a conditional statement. The fact that you can find a particular x such that $P(x) \Rightarrow Q(x)$ holds true does not mean that it is true for all x.

2.8 Proof by counterexample

To prove the negation of the statement $\forall x P(x)$, in other words to prove that $\exists x \, \neg P(x)$, we must find at least one x_0 such that $P(x_0)$ does not hold. Such an object is called a *counterexample* to $P(x)$.

Example 2.26. *Prove that the following statement is false: Every continuous function on an interval $[a, b]$ is differentiable on (a, b).*

Proof. For a counterexample, it suffices to consider $f(x) = |x|$ on the interval $[-1, 1]$ which is continuous, but not differentiable at $0 \in (-1, 1)$. □

Exercise 2.27. *Prove the negation of the following statements by giving counterexamples.*

    a. The sum of two irrational numbers is irrational.

    b. The product of two irrational numbers is irrational.

    c. If a and b are positive integers and $a \cdot b$ is a perfect square (there is an integer k with $a \cdot b = k^2$), then a and b are perfect squares.

    d. We have $\sqrt{a + b} = \sqrt{a} + \sqrt{b}$ for all $a, b \geq 0$.

2.9 Proof by mathematical induction

Recall that the principle of *mathematical induction* refers to the following: to prove that a sequence of statements $S(1), S(2), ..., S(n), ...$ is true, it suffices to prove two steps:

    Step 1. (Basis step) Prove that $S(1)$ is true,

    Step 2. (Inductive step) Assuming that $S(k)$ is true for a fixed integer $k \geq 1$, prove that $S(k + 1)$ is true.

    The mental image which helps us understand induction is the following: suppose we have a sequence of standing dominoes such that whenever we tip one domino, the next also falls. If the first domino is tipped, then all of them will fall eventually.

    The induction principle is based on the Peano axioms and the properties of positive integers, discussed in chapter 6. There are cases when $S(n)$ makes sense just for $n \geq n_0$, where n_0 is a fixed positive integer. In this case we first prove that $S(n_0)$ is true and assuming that $S(k)$ is true for a fixed $k \geq n_0$, we prove that $S(k + 1)$ is true. This is called generalized induction.

    The method of strong induction (or complete induction) assumes in step 2 that all $S(1), S(2), ..., S(k)$ are true and we prove that $S(k + 1)$ is true.

    We give now several examples of proofs by induction.

Example 2.28. *For all $n \geq 1$ we have $1 + 2 + \cdots + n = \dfrac{n(n + 1)}{2}$.*

Proof. We call $S(n)$ the statement to be proved. For $n = 1$ the statement becomes $1 = \dfrac{1 \cdot 2}{2}$, which is true.

    Assume $S(k)$ is true, i.e., $1 + \cdots + k = \dfrac{k(k + 1)}{2}$. Then

$$1 + \cdots + k + (k + 1) = \frac{k(k + 1)}{2} + k + 1 = \frac{(k + 1)(k + 2)}{2},$$

hence $S(k + 1)$ is also true.

Since both steps were verified, it follows by induction that $S(n)$ is true for all $n \geq 1$. $\qquad\square$

Example 2.29. *Suppose x is a real number with $x \neq 1$. Prove that for any positive integer n we have*

$$1 + x + \cdots + x^n = \frac{1 - x^{n+1}}{1 - x}.$$

Proof. The basis step is $1 + x = \dfrac{1 - x^2}{1 - x}$, which is true since $(1 + x)(1 - x) = 1 - x^2$.

Assume $1 + x + \cdots + x^k = \dfrac{1 - x^{k+1}}{1 - x}$ for a fixed $k \geq 1$. Adding x^{k+1} both sides we get

$$1 + x + \cdots + x^k + x^{k+1} = \frac{1 - x^{k+1}}{1 - x} + x^{k+1} = \frac{1 - x^{k+1} + x^{k+1}(1 - x)}{1 - x} =$$

$$= \frac{1 - x^{k+1} + x^{k+1} - x^{k+2}}{1 - x} = \frac{1 - x^{(k+1)+1}}{1 - x}.$$

$\qquad\square$

Example 2.30. *Show that $n^2 \leq 2^n$ for all $n \geq 4$.*

Proof. Note that $3^2 = 9 > 2^3 = 8$, so there is a good reason to start at $n_0 = 4$. For $n = 4$ the inequality becomes $4^2 = 16 \leq 2^4$ and the statement is true. Assume $k^2 \leq 2^k$ for some $k \geq 4$. We want to show that $(k+1)^2 \leq 2^{k+1}$. Since $2^{k+1} = 2 \cdot 2^k$, we multiply the inequality $k^2 \leq 2^k$ by 2 and we get $2k^2 \leq 2^{k+1}$. This is not yet what we wanted, but if we show that $(k+1)^2 \leq 2k^2$ for $k \geq 4$, we are done by putting together the two inequalities. Expanding the square, the last inequality is equivalent to $k^2 + 2k + 1 \leq 2k^2$ or $k^2 - 2k \geq 1$. If we write this as $k(k - 2) \geq 1$, we see that it is true, since $k \geq 4$. It follows that $n^2 \leq 2^n$ for all $n \geq 4$. $\qquad\square$

Example 2.31. *Prove that for each $n \geq 1$ we have $|\sin(nx)| \leq n \cdot \sin(x)$ for all $x \in [0, \pi]$.*

Proof. For $n = 1$ we need to show that $|\sin(x)| \leq \sin(x)$. Since $x \in [0, \pi]$, we have $\sin(x) \geq 0$, hence $|\sin(x)| = \sin(x)$ and the inequality is true.

Assume $|\sin(kx)| \leq k \sin(x)$ for a fixed $k \geq 1$ and $x \in [0, \pi]$. Then

$$|\sin((k + 1)x)| = |\sin(kx + x)| = |\sin(kx)\cos(x) + \cos(kx)\sin(x)| \leq$$

$$|\sin(kx)||\cos(x)| + |\cos(kx)||\sin(x)| \leq k\sin(x) + \sin(x) = (k + 1)\sin(x).$$

We used the fact that $\sin(a + b) = \sin(a)\cos(b) + \cos(a)\sin(b)$, the triangle inequality $|u + v| \leq |u| + |v|$, and the fact that $|\cos(kx)| \leq 1$ for all $k \geq 1$. The proof by induction is complete. $\qquad\square$

Example 2.32. *For any $n \geq 2$ prove that*

$$\frac{1}{n+1} + \frac{1}{n+2} + \cdots + \frac{1}{2n} \geq \frac{7}{12}.$$

Proof. For $n = n_0 = 2$ we have $\frac{1}{3} + \frac{1}{4} = \frac{7}{12}$, so the statement is true. It is easy to check that the inequality fails for $n = 1$. Assume

$$\frac{1}{k+1} + \frac{1}{k+2} + \cdots + \frac{1}{2k} \geq \frac{7}{12}$$

for some $k \geq 2$ and let's prove the inequality for $k + 1$. We have

$$\frac{1}{k+1+1} + \frac{1}{k+1+2} + \cdots + \frac{1}{2k} + \frac{1}{2k+1} + \frac{1}{2k+2} =$$

$$= \left(\frac{1}{k+1} + \frac{1}{k+2} + \cdots + \frac{1}{2k} \right) - \frac{1}{k+1} + \frac{1}{2k+1} + \frac{1}{2k+2}$$

and it suffices to notice that

$$-\frac{1}{k+1} + \frac{1}{2k+1} + \frac{1}{2k+2} = \frac{-2(2k+1) + (2k+2) + (2k+1)}{(2k+1)(2k+2)} =$$

$$= \frac{1}{(2k+1)(2k+2)} > 0.$$

□

Example 2.33. *For $n \geq 1$ prove that*

$$1 + \frac{1}{\sqrt{2}} + \frac{1}{\sqrt{3}} + \cdots + \frac{1}{\sqrt{n}} < 2\sqrt{n}.$$

Proof. For $n = 1$ the inequality becomes $1 < 2$, which is true. Assume

$$1 + \frac{1}{\sqrt{2}} + \frac{1}{\sqrt{3}} + \cdots + \frac{1}{\sqrt{k}} < 2\sqrt{k}$$

for some $k \geq 1$ and let's prove the inequality for $k + 1$. We have

$$1 + \frac{1}{\sqrt{2}} + \frac{1}{\sqrt{3}} + \cdots + \frac{1}{\sqrt{k}} + \frac{1}{\sqrt{k+1}} < 2\sqrt{k} + \frac{1}{\sqrt{k+1}} = \frac{2\sqrt{k^2+k} + 1}{\sqrt{k+1}} < 2\sqrt{k+1}$$

because $2\sqrt{k^2 + k} < 2k + 1$ and we are done.

□

Example 2.34. *Find the sum of the first n odd positive integers.*

Proof. This exercise has two parts: first we need to look for a pattern and guess the formula, and second we prove the formula by induction. We have

$$1 = 1 = 1^2, 1 + 3 = 4 = 2^2, 1 + 3 + 5 = 9 = 3^2, 1 + 3 + 5 + 7 = 16 = 4^2,$$

so a good guess for the sum of the first n odd integers looks like

$$1 + 3 + 5 + \cdots + (2n - 1) = n^2.$$

Note that there are n terms on the left-hand side. Obviously, we already know that this is true for $n = 1$ since $1 = 1^2$. Assume now that $1+3+\cdots+(2k-1) = k^2$ for a fixed arbitrary $k \geq 1$. Adding $2k + 1$ both sides, we get

$$1 + 3 + \cdots + (2k - 1) + (2k + 1) = k^2 + 2k + 1 = (k + 1)^2.$$

We conclude that the conjectured formula is true for any $n \geq 1$. $\square$

Example 2.35. *Suppose f is a function defined on the set of real numbers such that $f(x+y) = f(x) + f(y)$ for all x, y. Prove by induction that $f(nx) = nf(x)$ for any positive integer n and any real x.*

Proof. The statement is true for $n = 1$. Assume that $f(kx) = kf(x)$ and let's prove that $f((k + 1)x) = (k + 1)f(x)$. Indeed,

$$f((k + 1)x) = f(kx + x) = f(kx) + f(x) = kf(x) + f(x) = (k + 1)f(x).$$

$\square$

Exercise 2.36. *Here is a "proof" by induction that all horses have the same color. Find the gap in the "proof": Let $H(n)$ be the statement that any n horses have the same color. This is clearly true for $n = 1$, since one horse has a single color. Assume now that $H(k)$ is true and let's prove $H(k+1)$. Divide the $k+1$ horses in two groups: the horses $1, 2, ..., k$ and the horses $2, 3, ..., k + 1$. Since any k horses have the same color, each group has the same color. Since there are horses belonging to both groups, it follows that all $k + 1$ horses have the same color.*

Exercise 2.37. *Show by induction that the sum of the cubes of three consecutive natural numbers is divisible by 9.*

Proof. Indeed, $0^3 + 1^3 + 2^3 = 9$, which is divisible by 9. Assume

$$(k - 1)^3 + k^3 + (k + 1)^3 = 9m$$

for some $k, m \geq 1$ and let us prove that $k^3 + (k + 1)^3 + (k + 2)^3$ is a multiple of 9. Since

$$(k - 1)^3 = k^3 - 3k^2 + 3k - 1 \text{ and } (k + 2)^3 = k^3 + 6k^2 + 12k + 8,$$

we get

$$k^3 + (k + 1)^3 + (k + 2)^3 = 9m + 9k^2 + 3k + 9,$$

which is a multiple of 9. $\square$

2.10 Exercises

Exercise 2.38. *Use a direct proof to prove these statements.*
  a. *If x is odd, then x^3 is odd.*
  b. *Let a, b, c be integers. If $a \mid b$ and $a \mid c$, then $a \mid b + c$.*
  c. *If $x^2 + 5y = y^2 + 5x$, then $x = y$ or $x + y = 5$.*

Exercise 2.39. *Give a direct proof for the identity $\arcsin x + \arccos x = \dfrac{\pi}{2}$.*

Exercise 2.40. *Give direct proofs for the following implications and find counterexamples for the converse:*
  a) $(\forall x \, P(x)) \vee (\forall x \, Q(x)) \Rightarrow \forall x \, (P(x) \vee Q(x))$.
  b) $\exists x \, (P(x) \wedge Q(x)) \Rightarrow (\exists x \, P(x)) \wedge (\exists x \, Q(x))$.
  c) $\forall x \, (P(x) \Rightarrow Q(x)) \Rightarrow (\forall x \, P(x) \Rightarrow \forall x \, Q(x))$.
  d) $(\exists x \, P(x) \Rightarrow \exists x \, Q(x)) \Rightarrow \exists x (P(x) \Rightarrow Q(x))$.

Exercise 2.41. *Prove that the four midpoints of a quadrilateral are the vertices of a parallelogram.*

Exercise 2.42. *Consider a quadrilateral $ABCD$ in the plane. Denote by K, L, M, N the midpoints of the sides AB, BC, CD, DA respectively. Prove that the following are equivalent:*
  a) *The diagonals AC and BD are perpendicular.*
  b) $AB^2 + CD^2 = BC^2 + AD^2$.
  c) $KM = LN$.

Exercise 2.43. *Prove that $\sqrt{2} + \sqrt{6} < \sqrt{15}$ by contradiction.*

Exercise 2.44. *Prove that there are no rational numbers a and b such that $\sqrt{3} = a\sqrt{2} + b$.*

Exercise 2.45. *Let $a \neq 0$. Prove that the quadratic equation $ax^2 + bx + c = 0$ has distinct real roots if and only if $b^2 - 4ac > 0$.*

Exercise 2.46. *(requires Linear Algebra)*
  Suppose A is an $n \times n$ matrix with real entries. Prove that the following are equivalent.
  1) *A is invertible.*
  2) *The equation $A\mathbf{x} = \mathbf{b}$ has a unique solution for every $\mathbf{b} \in \mathbb{R}^n$.*
  3) *The equation $A\mathbf{x} = \mathbf{0}$ has only the trivial solution $\mathbf{x} = \mathbf{0}$.*
  4) *The reduced row echelon form of A is I_n, the identity matrix.*
  5) *$\det(A) \neq 0$.*
  6) *0 is not an eigenvalue for A.*

Exercise 2.47. *Show that the square of an integer ends in $0, 1, 4, 5, 6$ or 9.*

Exercise 2.48. *Solve the inequality $|x + 2| < |x^2 - 1|$ in $\mathbb{R}$ by considering several cases.*

Exercise 2.49. *Find the number of integers from 1 to 100 which have 6 as one of their digits.*

Exercise 2.50. *Prove that there is a positive integer n such that $2^n - 1$ is divisible by 11.*

Exercise 2.51. *Prove that there is a positive integer n such that $n^2 - n + 11$ is not prime.*

Exercise 2.52. *Let $f : (0, \infty) \to (0, \infty)$ be a function such that*

$$\forall\, x, y \in (0, \infty), \; f(x + y) = \frac{f(x)f(y)}{f(x) + f(y)}.$$

Prove by induction that $f(nx) = f(x)/n$ for all $n \geq 2$ and all $x > 0$.

Exercise 2.53. *Prove by induction that $1 + 5 + 9 + \cdots + (4n - 3) = 2n^2 - n$.*

Exercise 2.54. *Prove by induction that $1^2 + 3^2 + \cdots + (2n-1)^2 = (4n^3 - n)/3$.*

Exercise 2.55. *For $n \geq 1$, let $s_n = 1^2 + 2^2 + \cdots + n^2$.*
  a. Compute s_1, s_2, s_3, s_4 and try to conjecture a general formula for s_n.
  b. Prove that $s_n = \dfrac{n(n+1)(2n+1)}{6}$ by induction.

Exercise 2.56. *Consider the Fibonacci numbers f_n, where $f_1 = f_2 = 1$, $f_n = f_{n-2} + f_{n-1}$ for $n \geq 3$. Prove the following by induction:*
  a. $f_{n+1}^2 + 2f_n f_{n+1} = f_n f_{n+1} + f_{n+1} f_{n+2}$.
  b. $f_1^2 + f_2^2 + \cdots + f_n^2 = f_n f_{n+1}$.

Exercise 2.57. *For each positive integer n and any real number $x \geq -1$ prove by induction that $(1 + x)^n \geq 1 + nx$.*

Exercise 2.58. *Here is a "proof" by induction that any two positive integers are equal. Find the mistake:*
  For a, b positive integers, $\max(a, b)$ is defined to be the largest of a and b if $a \neq b$, and $\max(a, a) = a$. Let $P(n)$ be the statement: if a and b are positive integers such that $\max(a, b) = n$, then $a = b$. We use induction to prove that $P(n)$ is true for $n \geq 1$. For $n = 1$, since $\max(a, b) = 1$, we get $a = b = 1$. Assume $P(k)$ is true. Let a, b such that $\max(a, b) = k + 1$. Then $\max(a-1, b-1) = k$. Since we are assuming $P(k)$ is true, we get $a - 1 = b - 1$, hence $a = b$. Therefore $P(k + 1)$ is true, and by induction $P(n)$ is true for all $n \geq 1$. As a consequence, any two positive integers a, b are equal.

Exercise 2.59. *For $n \geq 10$ prove by generalized induction that $2^n \geq n^3$.*

Exercise 2.60. *Prove by induction that*

$$1 \cdot 2 + 2 \cdot 3 + 3 \cdot 4 + \cdots + n(n + 1) = \frac{n(n + 1)(n + 2)}{3}.$$

Exercise 2.61. *Prove by induction that for* $n \geq 1$

$$2 \cdot 6 \cdot 10 \cdots \cdot (4n - 2) = \frac{(2n)!}{n!}.$$

Recall that $n! = 1 \cdot 2 \cdot 3 \cdots n$.

Exercise 2.62. *Prove by induction that for* $n \geq 1$

$$(1 + 2^5 + \cdots + n^5) + (1 + 2^7 + \cdots + n^7) = 2 \left[\frac{n(n+1)}{2} \right]^4.$$

Exercise 2.63. *Prove by induction that the sum of internal angles in an* n-sided polygon $(n \geq 3)$ *is* $(n - 2)\pi$.

Exercise 2.64. *If we draw* n *straight lines in the plane, no three going through the same point, and no two parallel, how many regions do they determine in the plane? Prove by induction that the formula is* $(n^2 + n + 2)/2$.

Exercise 2.65. * *Some straight lines are drawn in the plane, forming regions. Show that it is possible to color each region either red or blue, in such a way that no two neighboring regions (regions separated by a line segment or a half line) have the same color.*

3

Elementary theory of sets

The notion of a set is a relatively recent development in mathematical history. Created at the end of 19th century, the idea of a set has become the cornerstone for virtually all contemporary mathematics. In its simplest form, a set is simply a bunch of things gathered together to form a new entity which is considered to be a single object. The English language is rich in words that could be regarded as synonymous with the word "set"; among them are collection, group, family, class, club, flock, herd, or team. Note that when we refer to a club, for example, we are not generally thinking of the individual members, but of the totality of all members, presumed to be a single entity. In fact, a club (or team, family, class, etc.) is often referred to in the singular.

We try to find a balance between a naive point of view of set theory and an axiomatic point of view. We only discuss the axiom of extent, which tells us when two sets are the same, and the axiom of separation, which enables us to form new sets from old ones. Later we will add the axiom of choice. To avoid paradoxes, like Russell's paradox, we will assume that our sets are subsets of a fixed big set named universe. We will talk about subsets of a given set, then we will define the power set and operations with sets like union, intersection, complement, difference and symmetric difference. Finally, we introduce ordered pairs and define the Cartesian product of sets.

3.1 Axioms for set theory

We will usually denote the sets by capital letters and we will mostly be interested in sets of mathematical entities like numbers, functions, etc. The fact that an element a belongs to a set A is written $a \in A$. The negation of $a \in A$ is written $a \notin A$. The sets can be specified by the roster notation using braces, like $A = \{a, b, c\}$, an enumeration of its elements, or by the set-builder notation, which will be explained below after the Axiom of Separation. When we enumerate the elements, a repetition should count only once. For example, the set $\{1, 2, 3, 1, 2\}$ is the same as $\{1, 2, 3\}$.

Example 3.1. *The following are examples of sets*

$$A = \{a, b, c, d, ..., z\}, B = \{\{1, 2, 5\}, +, -, *, A\},$$

$$C = \{2, 4, 6, ...\}, D = \{..., -2, -1, 0, 1, 2, ...\}.$$

Note that $10 \in C$ *but* $3 \notin C$. *The dots here indicate that the list continues indefinitely. You may recognize that* C *is the set of even positive integers, and* D *is the set of integers, also denoted by* $\mathbb{Z}$.

A set could be an element of another set. For example, notice that $\{1, 2, 5\} \in B$ *and* $A \in B$.

The main things that are needed in talking about sets are knowledge as to when sets are really the same and how to form sets. These are addressed in Axioms 1 and 2 below. Later, we will add a third axiom, called the *axiom of choice*. This will be explained in the next chapter.

When we use quantifiers, we will assume that the variables belong to a certain "universe". This universe can be understood like a large set for which our discussion makes sense.

Axiom 1 (The Axiom of Extent).

$$(A = B) \Leftrightarrow \forall x \, (x \in A \Leftrightarrow x \in B).$$

This axiom asserts that equality of sets is determined by the members of the sets and by no other criteria. Thus if A is described as the set of even integers and B is the set of all numbers that are obtained by increasing odd numbers by 1, then we are really dealing with the same set, even though the definitions of A and B are different.

In view of the Axiom of Extent, in order to prove that two sets are equal we must show two separate things, namely that each element of one set is also in the other and vice versa. This will usually be accomplished by breaking the proof into two parts. The first part will begin with a hypothesis of the form $x \in A$, from which the conclusion $x \in B$ will be drawn. The second part amounts to the converse of this. A proof of equality for sets is incomplete until the sentences $x \in A$ and $x \in B$ are shown to be equivalent.

Axiom 2 (The Axiom of Set Formation, or Axiom of Separation). If $S(x)$ is an open sentence in x (or predicate), then there is a set whose elements are exactly those x in the universe for which $S(x)$ is true. In symbols,

$$\exists A \, \forall x \, (x \in A \Leftrightarrow S(x)).$$

Because Axiom 2 allows formation of a set by collecting together all objects that make an open sentence true, you can see why the set formed is often called the *truth set* or *solution set* of the open sentence. Notice, however, that uniqueness of the solution set is not explicitly stated in Axiom 2.

As it happens, though, putting our two axioms together does show that the set formed in Axiom 2 is unique.

Theorem 3.2. $\exists! A \, \forall x \, (x \in A \Leftrightarrow S(x))$.

The fact that there is a set A satisfying the condition $x \in A \Leftrightarrow S(x)$

follows directly from Axiom 2. Now suppose A_1 and A_2 are any sets like A. Then we have

$$x \in A_1 \Leftrightarrow S(x) \Leftrightarrow x \in A_2.$$

Since the relation $\Leftrightarrow$ is transitive, in the sense that $P \Leftrightarrow Q$ and $Q \Leftrightarrow R$ implies $P \Leftrightarrow R$, we get $A_1 = A_2$ by Axiom 1.

In mathematics, when we know that something of interest exists and is unique, it is usually appropriate to give it a name. Thus we now introduce the standard notation for the truth set of an open sentence (defined in section 1.1).

Definition 3.3. *If $S(x)$ is an open sentence, the set A mentioned in the above theorem is denoted by $A = \{x : S(x)\}$. This is the set-builder notation (read: the set of all x such that $S(x)$).*

Remark 3.4. *The idea of set builder notation works in two ways. First, if you know that a is an element such that the sentence $S(a)$ is true, then it follows that a is an element of A. For example, since $2^3 - 2 = 6$, we know that $2 \in \{x : x^3 - x = 6\}$. Second, if you are presented with an element of A, then you know that the condition defining the set A is true. An example of this is that if you happen to run across an object, say b, which is definitely in $\{x : x^3 - x - 1 = 0\}$, even if you don't know a precise value for it, you can say with confidence that $b^3 - b - 1 = 0$.*

Here is another, more symbolic way, of expressing what was said above:

Theorem 3.5. $\forall a \, (a \in \{x : S(x)\} \Leftrightarrow S(a))$.

Example 3.6. *If we consider all positive integers n such that n is a multiple of 5, we get the set $A = \{5, 10, 15, 20, 25, ...\}$, in set builder notation*

$$A = \{n \text{ positive integer } : 5 \mid n\}.$$

3.2 Inclusion of sets

As we have seen, Axiom 1 deals with equality of sets. We define now the notion of subset, which illustrates the idea of a smaller set sitting inside a bigger set. More precisely, we have the following definition.

Definition 3.7. *We say that A is a subset of B or that A is contained in B whenever $\forall x (x \in A \Rightarrow x \in B)$. We write $A \subseteq B$. We may also express this by saying that B is a superset of A and write $B \supseteq A$. If $A \subseteq B$ and $A \neq B$, we say that A is a proper subset of B, and we write $A \subset B$.*

Example 3.8.
$$\{1, 2, 4\} \subseteq \{1, 2, 3, 4, 5\}, \ \mathbb{N} \subseteq \mathbb{Z}.$$

In fact,
$$\{1, 2, 4\} \subset \{1, 2, 3, 4, 5\}, \ \mathbb{N} \subset \mathbb{Z}.$$

Warning: some authors use $\subset$ for inclusion, and $\subsetneq$ for proper inclusion. Our choice of notation is parallel with inequality $\leq$ and strict inequality $<$ of numbers.

Note how the definition of subset resembles Axiom 1; the only difference is that the biconditional has been replaced by a single conditional. This means that being a subset of, is a weaker notion than being equal to. In fact, A will be a subset of B if and only if all the elements of A are also elements of B. Note that the converse may not be true.

Here are the fundamental properties of the notion of subset.

Theorem 3.9. *The inclusion of sets has these properties:*
  1. Reflexivity: For all A, $A \subseteq A$.
  2. Antisymmetry: $(A \subseteq B) \wedge (B \subseteq A) \Leftrightarrow A = B$.
  3. Transitivity: $(A \subseteq B) \wedge (B \subseteq C) \Rightarrow A \subseteq C$.

Proof. 1. This is clear, since $x \in A$ implies $x \in A$.

2. This is exactly the double implication used in Axiom 1 to conclude that two sets are equal.

3. Suppose $x \in A$. We can apply the subset definition to the first hypothesis and conclude that $x \in B$. But now this new assertion can be used with the second hypothesis to deduce that $x \in C$. This establishes the conditional $x \in A \Rightarrow x \in C$ for every x, so the result follows. □

Since Axiom 2 guarantees that there is a set associated with any open sentence, in particular we can define a set from a condition that can never be satisfied. And why not? After all, sets are often used to describe the solutions of various problems, and some problems have no solutions whatever. In the next definition we use a standard universally false statement to define the set we will refer to as the empty, null, or void set, denoted $\emptyset$.

Definition 3.10. *By definition, $\emptyset = \{x : x \neq x\}$.*

To understand the empty set, it might help if you consider the fact that a subset can be gotten from a set by removing some of its elements. If you take them all out, what do you get? Of course, you reach the empty set $\emptyset$.

Theorem 3.11. *We have*
  1. $\forall x \ (x \notin \emptyset)$.
  2. For all A, $\emptyset \subseteq A$.

Proof. 1. This is true since $\emptyset$ has no element.

2. The implication $x \in \emptyset \Rightarrow x \in A$ is true since the hypothesis is false. For another proof, suppose that the empty set is not a subset of a certain set A. That means that for some choice of x the sentence $x \in \emptyset \Rightarrow x \in A$ is false. This can only occur if, for this x, we have both $x \in \emptyset$ and $x \notin A$. As there is no element of the empty set, this is a contradiction. It must follow that $\emptyset \subseteq A$. □

Definition 3.12. *Given a set X, the power set $\mathcal{P}(X)$ is defined as*

$$\mathcal{P}(X) = \{A : A \subseteq X\},$$

the set of all subsets of X.

Note that $\emptyset, X \in \mathcal{P}(X)$ since $\emptyset \subseteq X$ and $X \subseteq X$.

Example 3.13. $\mathcal{P}(\{a, b, c\}) = \{\emptyset, \{a\}, \{b\}, \{c\}, \{a, b\}, \{a, c\}, \{b, c\}, \{a, b, c\}\}$ *has $2^3 = 8$ elements.*

Example 3.14. *Let us prove by induction that a set with n elements has 2^n subsets.*

Proof. The empty set $\emptyset$ has only one subset, namely itself. So the property holds for $n = 0$ since $2^0 = 1$. Assume that any set X with k elements has 2^k subsets. Let Y be a set with $k+1$ elements. We can write $Y = X \cup \{y\}$ where X has k elements and $y \notin X$. The subsets of Y are of two kinds: subsets which are included in X (there are 2^k of these) and subsets which contain y. The last type of subsets are of the form $A \cup \{y\}$ where $A \subseteq X$, hence a total of another 2^k. All together there are $2^k + 2^k = 2^{k+1}$ subsets of Y and we are done. $\qquad\square$

Obviously, some sets do not contain themselves. For example, let X be the set of integers. Then X is not an element of X, since X is not an integer. A set which contains itself must be a set containing sets as elements. Consider for example the set Y of sets which can be defined using 15 words or less. Clearly Y contains itself. Here is a surprising result, which will convince us that is best to avoid sets which contain themselves.

Theorem 3.15. *(Russell's paradox) The set $X = \{A : A \notin A\}$ is contradictory.*

Proof. Indeed, if the set X exists, then let's check if $X \in X$ or not. If we assume $X \in X$, then by definition $X \notin X$, a contradiction. If we assume $X \notin X$, then we conclude $X \in X$, a contradiction. $\qquad\square$

In particular, there is no set containing all sets. Such a thing is a new concept, usually called the class of all sets or the category of sets. We are not discussing category theory in this book. For those interested, there is an extensive literature on classes and category theory.

In the next sections, we will define several operations with sets, give examples, and prove the rules of algebra with sets. To avoid Russell's paradox, it is convenient to assume that all the sets we work with are subsets of a fixed large set U, called the universe. Many times, this set U will be understood from the context.

3.3   Union and intersection of sets

The two basic operations with sets are the *union* and the *intersection*. In the first, we gather together all those objects which appear in at least one of the sets. The second is obtained by putting together the common members of both sets into a single set. Recall that we assume our sets to be subsets of a fixed universe U.

Definition 3.16. *The union of the sets A and B is*

$$A \cup B = \{x \in U : x \in A \vee x \in B\}.$$

The intersection of A and B is the set

$$A \cap B = \{x \in U : x \in A \wedge x \in B\}.$$

Note especially the conditions written within braces that determine the qualifications required for an object to be a member. Sometimes we visualize these operations with pictures, called Venn diagrams. When we prove theorems about operations with sets, Venn diagrams are a useful visual aid.

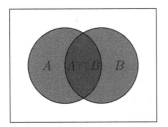 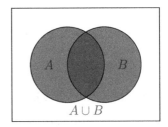

Definition 3.17. *Two sets A, B are called* disjoint *if $A \cap B = \emptyset$.*

Example 3.18. *Let $A = \{1, 2, c\}$, $B = \{a, b, c\}$ and $C = \{1, x\}$. Then*

$$A \cup B = \{1, 2, a, b, c\}, \quad A \cup C = \{1, 2, c, x\}, \quad A \cap B = \{c\},$$

$$A \cap C = \{1\}, \quad B \cup C = \{1, a, b, c, x\}, \quad B \cap C = \emptyset.$$

In particular, B and C are disjoint.

Example 3.19. *Consider the sets*

$$A = \{n \in \mathbb{Z} : n = 2k \text{ for some } k \in \mathbb{Z}\}, \quad B = \{n \in \mathbb{Z} : n = 3m \text{ for some } m \in \mathbb{Z}\}.$$

Then

$$A \cup B = \{n \in \mathbb{Z} : (n = 2k) \vee (n = 3m) \text{ for some } k, m \in \mathbb{Z}\},$$

$$A \cap B = \{n \in \mathbb{Z} : n = 6r \text{ for some } r \in \mathbb{Z}\}.$$

Here are some properties about intersection of sets.

Theorem 3.20. *For arbitrary sets A, B, C we have*
  1. $A \cap A = A$.
  2. $A \cap B = B \cap A$.
  3. $(A \cap B) \cap C = A \cap (B \cap C)$.

Proof. We must, in view of Axiom 1, prove the validity of a biconditional statement and, as we already mentioned, this is typically done by proving two conditional statements separately.

1. Suppose that $x \in A \cap A$. By definition, $x \in A$ and $x \in A$. Thus $x \in A$. This establishes the implication $x \in A \cap A \Rightarrow x \in A$, and one direction of the desired biconditional is established. Note the logical principle used here: whenever you have a statement of the form $P \wedge Q$ in a proof, you are entitled to simply write down P as a consequence. Of course, you may also properly infer the statement Q.

For the other direction of the biconditional, we now suppose that $x \in A$ and try to prove that $x \in A \cap A$. But this amounts to stating our hypothesis twice (which is surely valid; if a sentence P is true, then forming its conjunction with itself $P \wedge P$ will also produce a true sentence), and then using the definition of intersection.

The two steps used in the above proof are typical of the pattern used to prove equality of sets. You will generally divide the proof into two parts. In the first, introduce a symbol to stand for an arbitrary member of the left-hand set and proceed to show that it must also be in the right-hand set. Then do the reverse by considering a typical element of the right-hand set and proving that it must be in the other set. This technique will be called *proof by double inclusion*; see also part two in Theorem 3.9.

2. Suppose that $x \in A \cap B$. From the definition of intersection, we may infer that $x \in A$ and $x \in B$. But since asserting that two statements are both true can be done by mentioning either one first, it follows that $x \in B$ and $x \in A$ (see the commutative property in Theorem 1.9 in Chapter 1). Thus $x \in B \cap A$. This completes the first part of the proof, and normally we would write the details of the other direction. This time, however, the proof can be written by simply interchanging the letters A and B in the previous part, so it will be omitted.

3. The proof uses the fact that there is an associative type law for the word "and", see part a) in the same Theorem 1.9, and it will also be omitted. □

Exercise 3.21. *Prove that for sets A, B, C we have*
  1. $A \cup A = A$.
  2. $A \cup B = B \cup A$.
  3. $(A \cup B) \cup C = A \cup (B \cup C)$.

As you can see from the preceding results, union and intersection obey some of the more common laws of algebra (think addition and multiplication).

In the next theorem, you will see that there are some more similarities, but also some significant differences between number algebra and set algebra.

Theorem 3.22. *We have*

    1. $A \cap (A \cup B) = A$.

    2. $A \cap (B \cup C) = (A \cap B) \cup (A \cap C)$.

Proof. 1. Suppose $x \in A \cap (A \cup B)$. Then $x \in A$ and $x \in A \cup B$. In particular, $x \in A$, which means that $A \cap (A \cup B) \subseteq A$. On the other hand, assume $x \in A$. By definition of union, we have $x \in A \cup B$. Hence $x \in A \cap (A \cup B)$, and we get $A \subseteq A \cap (A \cup B)$, hence equality by part 2 of Theorem 3.9.

    2. Let us examine an element of $A \cap (B \cup C)$, call it x. Clearly $x \in A$ and $x \in B \cup C$. Since the last part of the preceding sentence implies that $x \in B$ or $x \in C$, we consider two cases.

    a) If $x \in B$, then since we already know that $x \in A$, we clearly have that $x \in A \cap B$. But then the statement

$$x \in A \cap B \vee x \in A \cap C$$

is also true. Thus, by definition of union,

$$x \in (A \cap B) \cup (A \cap C).$$

    b) If $x \in C$, then we have virtually the same argument as in part a). Check it out and see for yourself that this is so. Thus in any case, x appears in the right-hand set.

    Now choose an element x of $(A \cap B) \cup (A \cap C)$. Clearly $x \in A \cap B$ or $x \in A \cap C$. In either case, we have $x \in A$. However, in the first case x is in B, while in the second case x is an element of C; thus, no matter what, $x \in B \cup C$. Evidently we now know that $x \in A \cap (B \cup C)$, and the second part of the biconditional needed to apply Axiom 1 has been proved.

$$\square$$

The following theorem gives some relationships between the various concepts described so far. They are fairly useful in everyday reasoning about sets.

Theorem 3.23. *We have*

    1. $A \subseteq B \Leftrightarrow A \cap B = A$.

    2. $A \subseteq B \Leftrightarrow A \cup B = B$.

    3. $A \subseteq A \cup B$.

    4. $A \cap B \subseteq A$.

    5. $A \subseteq C \wedge B \subseteq C \Rightarrow A \cup B \subseteq C$.

    6. $A \subseteq B \wedge A \subseteq C \Rightarrow A \subseteq B \cap C$.

Proof. We will prove part 1 and leave the others as exercises. Notice first that the first statement is expressed as a biconditional, so there will be two parts to the proof. First we will assume that $A \subseteq B$ and use this to prove that

$A \cap B = A$. Then we will assume the truth of the equation and prove the inequality without referring to anything we did in the first part. Here goes!

i) Suppose $A \subseteq B$. Since we are trying to establish the equality of two sets, it behooves us to divide this portion of the proof into two parts.

a) Suppose $x \in A \cap B$. Then $x \in A$ and $x \in B$. Hence, $x \in A$.

b) Now suppose $x \in A$. Because we know that $A \subseteq B$, we can invoke the definition of inclusion to conclude $x \in B$. Thus $x \in A \cap B$.

Steps a) and b) together show that $x \in A \cap B \Leftrightarrow x \in A$, so by Axiom 1 we see that the sets are equal.

ii) Now assume that $A \cap B = A$. We wish to show that $A \subseteq B$. Let $x \in A$. Because A is exactly the same as $A \cap B$, we then know that $x \in A \cap B$. By definition of intersection, we have $x \in B$, and we are done. □

3.4 Complement, difference and symmetric difference of sets

Recall that all our sets are subsets of a universe U.

Definition 3.24. *The complement of the set A is $A' = \{x \in U : x \notin A\}$.*

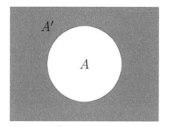

Another notation for the complement is A^c. The complement of A consists of all those objects in U which are not elements of A. For our purposes, the largest set U under consideration is usually understood, and complements are taken with respect to that set. Thus if we were studying the positive integers and we would consider $\{x : x < 5\}'$, then we would be referring to the set $\{5, 6, 7, ...\}$, whereas in a discussion of the set of digits in base eight $\{0, 1, 2, ..., 7\}$, then $\{x : x < 5\}'$ would be the set $\{5, 6, 7\}$.

The next theorem is the set-theoretic analogue of the double negation law of logic.

Theorem 3.25. *We have $(A')' = A$.*

Proof. $x \in (A')' \Leftrightarrow x \notin A' \Leftrightarrow \neg(x \in A') \Leftrightarrow \neg(\neg(x \in A)) \Leftrightarrow x \in A$.
By the Axiom of Extent, $(A')' = A$. □

We prove now other properties of the complement.

Theorem 3.26. *(De Morgan's Laws for complements) We have*
1. $(A \cap B)' = A' \cup B'$.
2. $(A \cup B)' = A' \cap B'$.

Proof. They are much like the laws of logic in Theorem 1.9 which show what happens when you negate a conjunction or disjunction. In fact, if you know that the sentence $\neg(P \wedge Q) \Leftrightarrow (\neg P) \vee (\neg Q)$ is a tautology (i.e., it is always true), then a simple proof of part 1 is:

$x \in (A \cap B)'$
$\Leftrightarrow \neg(x \in A \cap B)$
$\Leftrightarrow \neg(x \in A \wedge x \in B)$
$\Leftrightarrow \neg(x \in A) \vee \neg(x \in B)$
$\Leftrightarrow x \in A' \vee x \in B'$
$\Leftrightarrow x \in A' \cup B'$.

Let's prove part 2 by double inclusion. Suppose $x \in (A \cup B)'$. Then x is not an element of the set $A \cup B$. There doesn't seem to be any direct way to continue, but we can always ask and try to answer questions as we go. Right here a good question would be: "Is x a member of A"? The answer, naturally, is "no", because otherwise x would also be in $A \cup B$. In answering this question, we have discovered some useful information about x. Since the same process obviously leads to the conclusion that $x \notin B$, we can conclude that $x \in A' \wedge x \in B'$, so $x \in A' \cap B'$.

Now suppose that $x \in A' \cap B'$. Then x is not in A and also x is not in B. Could x be in the union of two sets if it is known that it is in neither? Of course not. Thus $x \notin (A \cup B)$, so by the definition of the complement, we get $x \in (A \cup B)'$.

$\square$

Definition 3.27. *The dual of a formula involving set variables, unions, and intersections, is the formula obtained by interchanging $\cup$ and $\cap$. The dual of an equation is the equation obtained by dualizing both sides.*

For example: the dual of $A \cup B$ is $A \cap B$, the dual of $B \cap (A \cup B)$ is $B \cup (A \cap B)$, and the dual of A is A itself. Looking back at some of the theorems of the algebra of sets, it is hard not to be struck by the fact that many of them appear in dual pairs. This is not an accident. To see why, let's look at the equation

$$A \cap (B \cup C) = (A \cap B) \cup (A \cap C).$$

We have already proved this to be true for all sets A, B, and C by a simple element argument. But if it is true for all sets, then we can replace the letters in it by symbols denoting any set and still obtain a valid statement. In particular, the following is true:

$$A' \cap (B' \cup C') = (A' \cap B') \cup (A' \cap C').$$

By taking complements of both sides and using De Morgan's Laws, we obtain

$$(A' \cap (B' \cup C'))' = ((A' \cap B') \cup (A' \cap C'))',$$

and

$$(A')' \cup ((B')' \cap (C')') = ((A')' \cup (B')') \cap ((A')' \cup (C')').$$

This clearly reduces to

$$A \cup (B \cap C) = (A \cup B) \cap (A \cup C).$$

In other words, we have a mechanical method for proving the dual of the original statement. This always works for every union-intersection identity. This fact is summarized in the following metatheorem.

Metatheorem. (Principle of Duality) The dual of any identity involving only set variables and union or intersection symbols is also an identity.

Definition 3.28. *Let X be a set and let $A, B \in \mathcal{P}(X)$. The difference of A and B, denoted by $A \setminus B$, is the set $\{x : x \in A \wedge x \notin B\}$.*

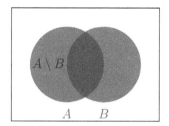

Example 3.29.

$$\{1, 2, 4, 5\} \setminus \{3, 4, 6\} = \{1, 2, 5\}, \quad \{3, 4, 6\} \setminus \{1, 2, 4, 5\} = \{3, 6\}.$$

Note that $A \setminus B$ may be different from $B \setminus A$. In particular we have $A' = X \setminus A$. Another notation for the set difference is $A - B$. We prefer $A \setminus B$ since, for A and B sets of numbers,

$$A - B = \{a - b : a \in A, b \in B\}$$

has a different meaning. For example,

$$\{1, 2, 3\} - \{1\} = \{1 - 1, 2 - 1, 3 - 1\} = \{0, 1, 2\}.$$

An object is in $A \setminus B$ if and only if it is in A and it is not in B. Looking at the other side of the coin, how will an element fail to be in $A \setminus B$? Clearly there are two ways: 1) if $x \notin A$, then x doesn't satisfy the first requirement, so x is not an element of $A \setminus B$; or 2) if $x \in B$, then x is not in the difference because the second part of the definition is wrong. Thus we have the equivalence $x \notin A \setminus B \Leftrightarrow [x \notin A \vee x \in B]$.

Theorem 3.30. *For arbitrary sets A, B, C we have*

    1. $A \setminus B = A \cap B'$.
    2. $(A \cup B) \setminus C = (A \setminus C) \cup (B \setminus C)$.
    3. $A \setminus (B \setminus C) = (A \setminus B) \cup (A \cap C)$.
    4. $A \setminus (B \cup C) = (A \setminus B) \setminus C$.

Proof. 1. This is true by definition.

    2. $(A \cup B) \setminus C = (A \cup B) \cap C' = (A \cap C') \cup (B \cap C') = (A \setminus C) \cup (B \setminus C)$.

    3. $A \setminus (B \setminus C) = A \cap (B \cap C')' = A \cap (B' \cup C) = (A \cap B') \cup (A \cap C) = (A \setminus B) \cup (A \cap C)$.

    4. We have $A \setminus (B \cup C) = A \cap (B \cup C)' = A \cap B' \cap C'$ and $(A \setminus B) \setminus C = (A \cap B') \cap C' = A \cap B' \cap C'$, so the left-hand side is equal to the right-hand side. $\qquad\square$

Remark 3.31. *Let A_1, A_2 be arbitrary sets. Then there are disjoint sets B_1, B_2 such that*

$$A_1 \cup A_2 = B_1 \cup B_2.$$

(Recall that X, Y are disjoint if $X \cap Y = \emptyset$).

Proof. Take $B_1 = A_1, B_2 = A_2 \setminus A_1$. $\qquad\square$

Definition 3.32. *The symmetric difference of two sets A and B is defined as*

$$A \Delta B = (A \setminus B) \cup (B \setminus A).$$

Another notation for the symmetric difference is $A \oplus B$.

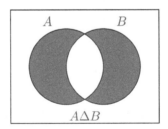

Example 3.33. *Let $A = \{a, b, c, d\}, B = \{c, d, e, f\}$. Then*

$$A \Delta B = \{a, b, e, f\}.$$

Theorem 3.34. *The symmetric difference has the following properties:*

    1. $A \Delta B = B \Delta A$ (commutativity).
    2. $(A \Delta B) \Delta C = A \Delta (B \Delta C)$ (associativity).
    3. $A \Delta \emptyset = \emptyset \Delta A = A$ and $A \Delta A = \emptyset$.
    4. $A \Delta B = C \Leftrightarrow A \Delta C = B$.
    5. $A \cap (B \Delta C) = (A \cap B) \Delta (A \cap C)$ (distributivity).

Proof. 1. $A\triangle B = (A \setminus B) \cup (B \setminus A) = (B \setminus A) \cup (A \setminus B) = B\triangle A$.

2. We have

$$(A\triangle B)\triangle C = [((A \cap B') \cup (B \cap A')) \cap C'] \cup [C \cap ((A \cap B') \cup (B \cap A'))'] =$$

$$(A \cap B' \cap C') \cup (B \cap A' \cap C') \cup [C \cap ((A \cap B')' \cap (B \cap A')')] =$$

$$(A \cap B' \cap C') \cup (B \cap A' \cap C') \cup [C \cap ((A' \cup B) \cap (B' \cup A))] =$$

$$(A\cap B'\cap C')\cup(B\cap A'\cap C')\cup[C\cap((A'\cap B')\cup(B\cap B')\cup(A'\cap A)\cup(B\cap A))] =$$

$$(A \cap B' \cap C') \cup (B \cap A' \cap C') \cup (C \cap A' \cap B') \cup (C \cap B \cap A)$$

since $B \cap B' = A' \cap A = \emptyset$. Notice that the expression for $(A\triangle B)\triangle C$ is symmetric in A, B, C. Starting with the right-hand side $A\triangle(B\triangle C)$, we get the same expression (convince yourself!).

3. It follows from the definition.

4. Assume $A\triangle B = C$. Taking the symmetric difference with A we get $A\triangle(A\triangle B) = A\triangle C$. Using associativity, $(A\triangle A)\triangle B = A\triangle C$. Since $A\triangle A = \emptyset$ and $\emptyset\triangle B = B$, we get $B = A\triangle C$. The converse is proved similarly (exercise!).

5. $A\cap(B\triangle C) = A\cap((B\cap C')\cup(B'\cap C)) = (A\cap B\cap C')\cup(A\cap B'\cap C)$. On the other hand, $(A\cap B)\triangle(A\cap C) = ((A\cap B)\cap(A\cap C)')\cup((A\cap B)'\cap(A\cap C)) = (A\cap B\cap A')\cup(A\cap B\cap C')\cup(A'\cap A\cap C)\cup(B'\cap A\cap C) = (A\cap B\cap C')\cup(B'\cap A\cap C)$. $\square$

Other properties related to the symmetric difference are contained in

Exercise 3.35. *We have*

a. $A\triangle B = (A \cup B) \setminus (A \cap B)$.
b. $A \cup B = (A\triangle B) \cup (A \cap B)$.
c. $A \setminus B = A\triangle(A \cap B)$.

Example 3.36. *(Equations with sets). Let's find sets X and Y satisfying all these conditions:*

a. $X \cup Y = \{1, 2, 3, 4, 5, 6, 7, 8, 9\}$.
b. $X \cap Y = \{4, 6, 9\}$.
c. $X \cup \{3, 4, 5\} = \{1, 3, 4, 5, 6, 8, 9\}$.
d. $Y \cup \{2, 4, 8\} = \{2, 4, 5, 6, 7, 8, 9\}$.

Solution. *We know that*

$$\{4, 6, 9\} \subseteq X, Y \subseteq \{1, 2, 3, 4, 5, 6, 7, 8, 9\}.$$

From part c it follows that $1, 8 \in X$ and from d we get that $5, 7 \in Y$. Now $2, 3 \in X\cup Y$, so they belong to one of the sets, but not to both since $2, 3 \notin X\cap Y$. It follows that there are several solutions:

$$X_1 = \{1, 2, 3, 4, 6, 8, 9\}, Y_1 = \{4, 5, 6, 7, 9\}.$$

$$X_2 = \{1, 2, 4, 6, 8, 9\}, Y_2 = \{3, 4, 5, 6, 7, 9\}.$$

$$X_3 = \{1, 3, 4, 6, 8, 9\}, Y_3 = \{2, 4, 5, 6, 7, 9\}.$$

$$X_4 = \{1, 4, 6, 8, 9\}, Y_4 = \{2, 3, 4, 5, 6, 7, 9\}.$$

3.5   Ordered pairs and the Cartesian product

An important operation with sets, analogue to the multiplication of numbers, is the Cartesian product $A \times B$. It consists of all ordered pairs $\langle a, b \rangle$ with $a \in A$ and $b \in B$.

Definition 3.37. *The ordered pair of x and y, denoted by $\langle x, y \rangle$, is the set* $\{\{x\}, \{x, y\}\}$.

Many books will use (x, y) for the ordered pair. We prefer $\langle x, y \rangle$ over (x, y) because of the conflict with the open interval notation (a, b) for $a, b \in \mathbb{R}$.

Theorem 3.38. $\langle y, z \rangle = \langle u, v \rangle \Leftrightarrow y = u$ *and* $z = v$.

Proof. Assuming $\langle y, z \rangle = \langle u, v \rangle$, we get $\{\{y\}, \{y, z\}\} = \{\{u\}, \{u, v\}\}$. It follows that either $\{y\} = \{u\}$ and $\{y, z\} = \{u, v\}$, or $\{y\} = \{u, v\}$ and $\{y, z\} = \{u\}$. In the first case it follows that $y = u$ and $z = v$. In the second case we get $u = v = y = z$. The converse is trivial. $\qquad\square$

Definition 3.39. *The Cartesian product of two sets X and Y is*

$$X \times Y = \{\langle x, y \rangle : x \in X, y \in Y\}$$

i.e., the set of all ordered pairs with the first component taken from X and the second component taken from Y.

Example 3.40. *If $X = \{1, 2, 3\}$ and $Y = \{a, b\}$, then*

$$X \times Y = \{\langle 1, a \rangle, \langle 2, a \rangle, \langle 3, a \rangle, \langle 1, b \rangle, \langle 2, b \rangle, \langle 3, b \rangle\}.$$

In calculus, you probably already used $\mathbb{R}^2$ for $\mathbb{R} \times \mathbb{R}$, the set of pairs of real numbers, identified with the points in the xy-plane, or $\mathbb{R}^3 = \mathbb{R} \times \mathbb{R} \times \mathbb{R}$, the set of triples of real numbers, identified with the three-dimensional space.

Example 3.41. *The Cartesian product of intervals $[0, 1] \times [1, 3]$ can be visualized as a rectangle in the plane $\mathbb{R}^2$:*

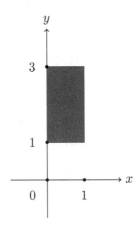

Example 3.42. *Let's find $X \times Y$ if $X = \{x \in \mathbb{Z} : x^2 = 16\}$ and $Y = \{y \in \mathbb{R} : |y - 1| \leq 5\}$.*

Solution. We have $X = \{-4, 4\}$ and $Y = [-4, 6]$. It follows that

$$X \times Y = \{\langle x, y \rangle : x = \pm 4, \ y \in [-4, 6]\}.$$

We can visualize this Cartesian product as a union of two segments in the plane:

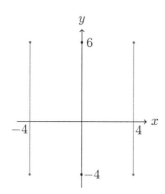

Here are some properties of the Cartesian product in regards to other set operations:

Theorem 3.43. *We have*
   1. $(A \cup B) \times C = (A \times C) \cup (B \times C)$.
   2. $(A \cap B) \times C = (A \times C) \cap (B \times C)$.
   3. $(A \setminus B) \times C = (A \times C) \setminus (B \times C)$.

Proof. 1. We have $\langle x, y \rangle \in (A \cup B) \times C$ equivalent to $x \in A \cup B$ and $y \in C$. This is the same as $x \in A$ and $y \in C$ or $x \in B$ and $y \in C$, in other words, $\langle x, y \rangle \in A \times C$ or $\langle x, y \rangle \in B \times C$, which means $\langle x, y \rangle \in (A \times C) \cup (B \times C)$.
   2. The proof is similar to the proof of part 1.
   3. We have $\langle x, y \rangle \in (A \setminus B) \times C \Leftrightarrow x \in A \setminus B \wedge y \in C \Leftrightarrow x \in (A \cap B') \wedge y \in C \Leftrightarrow x \in A \wedge x \notin B \wedge y \in C \Leftrightarrow \langle x, y \rangle \in A \times C \wedge \langle x, y \rangle \notin B \times C \Leftrightarrow \langle x, y \rangle \in (A \times C) \setminus (B \times C)$. $\square$

Definition 3.44. *The disjoint union of two sets X, Y, denoted $X \sqcup Y$, is defined as $(X \times \{1\}) \cup (Y \times \{2\})$.*

The idea is that if $X \cap Y \neq \emptyset$, then the elements in the intersection need new labels to be distinguished. For example,

$$\{a, b, c\} \sqcup \{b, c, d\} = \{\langle a, 1 \rangle, \langle b, 1 \rangle, \langle c, 1 \rangle, \langle b, 2 \rangle, \langle c, 2 \rangle, \langle d, 2 \rangle\}.$$

The sets X, Y can be identified with the subsets $X \times \{1\}, Y \times \{2\}$ of $X \sqcup Y$.

3.6   Exercises

Note. If not otherwise specified, the sets are assumed to be subsets of a fixed universe.

Exercise 3.45. *List all the elements in the set* $A = \{x \in \mathbb{Z} : |x - 3| \leq 5\}$.

Exercise 3.46. *Let* $A = \{x \in \mathbb{R} : |x - 1| < 1\}$ *and let* $B = \{x \in \mathbb{R} : |x - 2| < 2\}$. *Prove that* $A \subseteq B$.

Exercise 3.47. *Let* $C = \{x \in \mathbb{R} : |x - 3| < 1\}$ *and let* $D = \{x \in \mathbb{R} : |x - 6| < 2\}$. *Prove that the sets* C *and* D *are disjoint.*

Exercise 3.48. *Find* $\mathcal{P}(\{1, 2, 3, 4\})$.

Exercise 3.49. *Prove that*
  1. $A \cup (A \cap B) = A$.
  2. $A \cup (B \cap C) = (A \cup B) \cap (A \cup C)$.

Exercise 3.50. *We have*
  a. $A \cap \emptyset = \emptyset$.
  b. $A \cup \emptyset = A$.

Exercise 3.51. *Let* $A, B \subseteq X$. *Show that*
  1. $A \subseteq B \Leftrightarrow B' \subseteq A'$.
  2. $A \cap A' = \emptyset$.
  3. $\forall x \in X\ (x \in A \cup A')$.

Exercise 3.52. *Let* A, B, C *be three sets such that*

$$A \cup B = C, \quad (A \cup C) \cap B = C, \quad (A \cap C) \cup B = A.$$

Prove that $A = B = C$.

Exercise 3.53. *Show that* $(A \setminus B)' = A' \cup B$.

Exercise 3.54. *For sets* A, B, C, *prove that* $A \setminus (B \cap C) = (A \setminus B) \cup (A \setminus C)$.

Exercise 3.55. *Let* $A_1, A_2, ..., A_n$ *be arbitrary sets for* $n \geq 2$. *Prove that there are disjoint sets* $B_1, B_2, ..., B_n$ *such that*

$$A_1 \cup A_2 \cup \cdots \cup A_n = B_1 \cup B_2 \cup \cdots \cup B_n.$$

Exercise 3.56. *Find sets* X *and* Y *satisfying all these conditions:*
  a. $X \cup Y = \{1, 2, 3, 4, 5, 6\}$.
  b. $X \cap Y = \{1, 2, 3, 4\}$.
  c. $\{4, 6\}$ *is not a subset of* X.
  d. $\{5, 6\}$ *is not a subset of* $Y \setminus X$.

Exercise 3.57. *Find all sets X and Y satisfying $X \Delta Y = \{1,2,3,4\}$ and $X \cap Y = \{5,6\}$. How many solutions do we have?*

Exercise 3.58. *Solve each of these equations for X:*
  a. $A \cup (B \setminus X) = B \cup X$ if $A = \{1,2,3\}, B = \{3,4,5\}$;
  b. $\{1,2\} \Delta X = \{1,2,3\}$;
  c. $(\{1,2\} \Delta X) \Delta \{1,2,3\} = \{1,2,3,4\}$.

Exercise 3.59. *Let X and Y be nonempty sets. Prove that $X \times Y = Y \times X$ if and only if $X = Y$.*

Exercise 3.60. *Prove the following:*
  a. $(A \cup B) \times (C \cup D) = (A \times C) \cup (B \times C) \cup (A \times D) \cup (B \times D)$.
  b. $(A \cap B) \times (C \cap D) = (A \times C) \cap (B \times C) \cap (A \times D) \cap (B \times D)$.

4

Functions

Many times we deal with quantities which depend on other quantities: the volume depends on the size, the heat index depends on the humidity, the force depends on the mass. These illustrate the idea of a function. In mathematics, a function is a certain rule which associates to any element in a set A, called the domain, a unique element in a set B, called the codomain. You already met real functions of real variables in calculus, like $f(x) = x^2, g(x) = \ln x$ or $h(x) = \tan x$. The domain and the set of values for these functions are subsets of $\mathbb{R}$, and a function is defined as a formula (or algorithm) which associates to each input in the domain a precise output. The domain of f is $\mathbb{R}$ and the set of values is $[0, \infty)$. The domain of g is $(0, \infty)$ and the set of values is $\mathbb{R}$. The domain of h is $\mathbb{R} \setminus \{(2k+1)\frac{\pi}{2} : k \in \mathbb{Z}\}$ and the set of values is $\mathbb{R}$. We will need to work with more general functions among all kinds of sets, not just subsets of the real numbers.

Even though a function is a particular case of a relation, we study functions first and define relations in the next chapter. After giving the precise definition of a function using its graph, we introduce operations and give several examples of functions. A given function determines two new functions, called the direct image and the inverse image, where the inputs and the outputs are sets. Sometimes it is necessary to shrink or enlarge the domain of a function, giving rise to restrictions and extensions. We also discuss one-to-one and onto functions, composition, and inverse functions. We conclude with families of sets and the axiom of choice, necessary in many proofs.

4.1 Definition and examples of functions

Here is the formal definition of a function.

Definition 4.1. *A function from a set X to a set Y is a subset f of the Cartesian product $X \times Y$ such that for all $x \in X$ there is a unique $y \in Y$ with $\langle x, y \rangle \in f$.*

The set X is called the *domain* of f, denoted $\text{dom}(f)$, and the set

$$\{y \in Y : \exists\, x \in X \text{ with } \langle x, y \rangle \in f\}$$

is called the *range* of f, denoted ran(f). The set Y is the set where f takes values, also called the *codomain* of f. Note that the range ran(f) may be a proper subset of the codomain Y. We write $f : X \rightarrow Y$, and for each $x \in X$ the unique element $y \in Y$ such that $\langle x, y \rangle \in f$ is denoted $f(x)$.

The set of all ordered pairs $\langle x, f(x) \rangle$ for $x \in X$ is also called the *graph* of f. Note that in our definition of a function, f, is determined by its graph, which is a subset of $X \times Y$.

Two functions will be equal if and only if they have the same domain, the same codomain, and the same graph.

Example 4.2. *Let $X = Y = \mathbb{R}$ and let $f : X \rightarrow Y, f(x) = x^2$. Here dom($f$) = $\mathbb{R}$, ran(f) = $[0, \infty)$ and the graph of f is the familiar parabola in the xy-plane.*

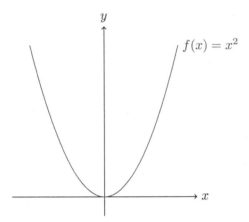

Example 4.3. *Let $X = \{1, 2, 3\}, Y = \{a, b, c\}$ and $f = \{\langle 1, c \rangle, \langle 2, a \rangle, \langle 3, a \rangle\}$. Then f is a function from X to Y such that $f(1) = c, f(2) = a, f(3) = a$ and ran(f) = $\{a, c\}$.*

We can visualize this function in a diagram:

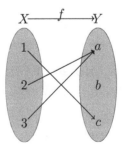

Notice that we join, by an arrow, each element in the domain with the corresponding element in the codomain.

Example 4.4. *Let X be any set. We define the* identity *function* id_X *by* $id_X : X \to X, id_X(x) = x$. *Its domain and range are both X and the graph of id_X is the diagonal*

$$\{\langle x, x \rangle : x \in X\}.$$

If $X = [0, \infty)$, then the graph looks like this:

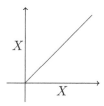

Example 4.5. *Let X, Y be any nonempty sets and let $c \in Y$. Then $f : X \to Y, f(x) = c$ for all $x \in X$ is called a* constant *function.*

Example 4.6. *As we already mentioned, in calculus a function was specified by a formula, like $f(x) = \sqrt{x}$, with the understanding that the domain is the largest set of real numbers for which $f(x)$ makes sense. In this case, $dom(f) = [0, \infty)$, $ran(f) = [0, \infty)$ and we can write*

$$f : [0, \infty) \to \mathbb{R}, f(x) = \sqrt{x}$$

or

$$f : [0, \infty) \to Y, f(x) = \sqrt{x},$$

where Y is any set such that $[0, \infty) \subseteq Y \subseteq \mathbb{R}$. By changing Y we get different functions, since they have different codomains. The graph of $f : [0, \infty) \to \mathbb{R}, f(x) = \sqrt{x}$ is a subset of $[0, \infty) \times \mathbb{R}$:

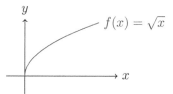

Exercise 4.7. *Explain what is wrong with the following "functions":*

   a. Let $f : \mathbb{R} \to \mathbb{R}, f(x) = \dfrac{1}{x^2 - 1}$.
   b. Let $g : [0, 5] \to [0, 2], g(x) = x - 1$.
   c. Let $h : [-1, \infty) \to [3, 4), h(x) = \sqrt{x - 1}$.

Solution. a. Note that x cannot be ± 1, since these values vanish the denominator. The correct domain of f should be a subset of $\mathbb{R} \setminus \{\pm 1\} = (-\infty, -1) \cup (-1, 1) \cup (1, \infty)$.

b. When $x \in [0,5]$, $g(x) = x - 1 \in [-1,4]$, so the correct codomain of g is $[-1,4]$ or any set containing this interval.

c. The square root $\sqrt{x-1}$ is defined only for $x \geq 1$ and it takes values in $[0,\infty)$. The correct domain of h is any subset of $[1,\infty)$ and the correct codomain is $[0,\infty)$ or any set containing this interval.

Example 4.8. *(Integer part) For any real number x, denote by $\lfloor x \rfloor$ the largest integer k such that $k \leq x$. For example, $\lfloor 1.2 \rfloor = 1$, $\lfloor -5.3 \rfloor = -6$. Then $f : \mathbb{R} \to \mathbb{Z}$, $f(x) = \lfloor x \rfloor$ is a function, called the integer part function or the floor function, with the following graph.*

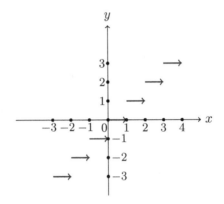

There is also a *ceiling* function, denoted $\lceil x \rceil$, defined as the smallest integer larger than or equal to x.

Example 4.9. *(Fractional part) Let $f : \mathbb{R} \to [0,1)$, $f(x) = x - \lfloor x \rfloor$. Then f is called the fractional part function. It has the following graph:*

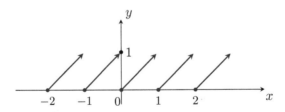

Exercise 4.10. *Prove that $x - 1 < \lfloor x \rfloor \leq x$. Use this inequality to show that for $x \geq 1$ we have*

$$\frac{1}{2} < \frac{\lfloor x \rfloor}{x} \leq 1.$$

Exercise 4.11. *Graph the ceiling function $f : \mathbb{R} \to \mathbb{Z}$, $f(x) = \lceil x \rceil$.*

Example 4.12. *(Characteristic function) For X any set and $A \subseteq X$, denote by χ_A the function*

$$\chi_A : X \to \{0,1\}, \quad \chi_A(x) = \begin{cases} 1 & \text{if } x \in A \\ 0 & \text{if } x \notin A. \end{cases}$$

This is called the characteristic *(or indicator) function of A.*

Definition 4.13. *(Operations with functions) Given two functions f, g with real values, define $f + g, f - g, f \cdot g$, and f/g to be new functions such that*

$$(f + g)(x) = f(x) + g(x), \quad (f - g)(x) = f(x) - g(x),$$

$$(f \cdot g)(x) = f(x) \cdot g(x), \quad (f/g)(x) = f(x)/g(x),$$

whenever this makes sense.

The characteristic function has the following properties:

Theorem 4.14. *Consider a set X and $A, B \in \mathcal{P}(X)$. Then*
  a. $A = B$ iff $\chi_A = \chi_B$.
  b. $\chi_{A \cap B} = \chi_A \cdot \chi_B$.
  c. $\chi_{A \cup B} = \chi_A + \chi_B - \chi_A \cdot \chi_B$.
  d. $\chi_{A'} = 1 - \chi_A$, where 1 denotes the constant function equal to 1 for all $x \in X$.
  e. $\chi_{A \setminus B} = \chi_A - \chi_A \cdot \chi_B$.
  f. $\chi_{A \Delta B} = \chi_A + \chi_B - 2\chi_A \cdot \chi_B$.

Proof. a. If $A = B$, then it is clear that $\chi_A, \chi_B : X \to \{0,1\}$ have the same values, so $\chi_A = \chi_B$. Conversely, assume $\chi_A = \chi_B$ and suppose $x \in A$. This happens if and only if $\chi_A(x) = 1 = \chi_B(x)$, i.e., $x \in B$. By double inclusion we get $A = B$.

b. We can directly check that $\chi_A(x)\chi_B(x) = 1$ precisely when $\chi_A(x) = \chi_B(x) = 1$ i.e. when $x \in A \cap B$, and otherwise $\chi_A(x)\chi_B(x) = 0$.

c. We have $\chi_{A \cup B}(x) = 1$ when $x \in A$ or $x \in B$, and otherwise $\chi_{A \cup B}(x) = 0$. Now $\chi_A(x) + \chi_B(x) = 2$ precisely when $x \in A \cap B$. It follows that in any case $\chi_{A \cup B} = \chi_A + \chi_B - \chi_A \cdot \chi_B$.

d. We have $x \in A'$ precisely when $x \notin A$, so $\chi_{A'}(x) = 1$ when $\chi_A(x) = 0$ and $\chi_{A'}(x) = 0$ when $\chi_A(x) = 1$. It follows that $\chi_{A'} = 1 - \chi_A$.

e. We have $A \setminus B = A \cap B'$ and we can apply properties b and d.

f. We have $A \Delta B = (A \cup B) \setminus (A \cap B)$, so

$$\chi_{A \Delta B} = \chi_{A \cup B} - \chi_{A \cup B}\chi_{A \cap B} = \chi_A + \chi_B - \chi_A\chi_B - (\chi_A + \chi_B - \chi_A\chi_B)\chi_A\chi_B =$$

$$\chi_A + \chi_B - \chi_A\chi_B - \chi_A^2\chi_B + \chi_A\chi_B^2 - \chi_A^2\chi_B^2 = \chi_A + \chi_B - 2\chi_A \cdot \chi_B$$

since $\chi_A^2 = \chi_A, \chi_B^2 = \chi_B$. □

Exercise 4.15. *Use the above properties of the characteristic function to prove that $A \Delta (B \Delta C) = (A \Delta B) \Delta C$ for arbitrary sets $A, B, C \in \mathcal{P}(X)$.*

Definition 4.16. *For $x, y \in \mathbb{R}$ define the maximum and the minimum functions by*

$$\max(x, y) = \begin{cases} x & \text{if} \quad x \geq y \\ y & \text{if} \quad x < y \end{cases}$$

and

$$\min(x, y) = \begin{cases} x & \text{if} \quad x \leq y \\ y & \text{if} \quad x > y, \end{cases}$$

respectively. Here the domain of both functions is $\mathbb{R} \times \mathbb{R}$ and the range is $\mathbb{R}$.

Notice that the notation $\max(\langle x, y \rangle)$ would be awkward. Some people use the notation $\max\{x, y\}$ and $\min\{x, y\}$ for the same functions.

Example 4.17. *The* sign *function is $sgn : \mathbb{R} \to \{-1, 0, 1\}$ such that*

$$sgn(x) = \begin{cases} -1 & \text{if} \quad x < 0 \\ 0 & \text{if} \quad x = 0 \\ 1 & \text{if} \quad x > 0. \end{cases}$$

Exercise 4.18. *Graph the* sign *function.*

Sometimes a function $f : X \to Y$ can be defined using a certain property satisfied by ordered pairs $\langle x, y \rangle$. More precisely, we have

Theorem 4.19. *Suppose X, Y are sets and suppose $P(x, y)$ is an open sentence depending on $\langle x, y \rangle \in X \times Y$ such that $\forall x \in X \; \exists! y \in Y$ such that $P(x, y)$. Then $\{\langle x, y \rangle : x \in X \wedge P(x, y)\}$ is a function with domain X and codomain Y.*

Proof. Since for all $x \in X$ there is a unique $y \in Y$ satisfying $P(x, y)$, by taking $f(x) = y$ we get a function $f : X \to Y$. $\square$

Example 4.20. *Consider $P(x, y)$ to be $2x + y = 1$ for $x, y \in \mathbb{Z}$. Then, since we can solve for y and there is a unique solution $y = 1 - 2x$, we can define the function $f : \mathbb{Z} \to \mathbb{Z}, f(x) = 1 - 2x$.*

4.2 Direct image, inverse image

Definition 4.21. *(Direct image function) Given $f : X \to Y$, we define a new function $f_{\mathcal{P}} : \mathcal{P}(X) \to \mathcal{P}(Y)$ such that*

$$f_{\mathcal{P}}(A) = \{f(a) : a \in A\}.$$

Many times we drop the subscript $\mathcal{P}$ and denote the new function also by f, even though this is now a function of sets. The set $f(A)$ is called the direct image of A. In particular note that $f(X) = \text{ran}(f)$. It should be clear from the context if we talk about $f : X \to Y$ or about $f : \mathcal{P}(X) \to \mathcal{P}(Y)$.

Example 4.22. *If* $X = \{a, b, c\}$, $Y = \{1, 2, 3\}$ *and* $f(a) = 2, f(b) = f(c) = 1$, *then* $f(\emptyset) = \emptyset, f(\{a\}) = \{2\}, f(\{b\}) = f(\{c\}) = \{1\}, f(\{a, b\}) = f(\{a, c\}) = \{1, 2\}, f(\{b, c\}) = \{1\}, f(X) = \{1, 2\}$.

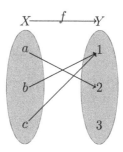

Definition 4.23. *(Inverse image function) Given* $f : X \rightarrow Y$, *we define* $f_{\mathcal{P}}^{-1} : \mathcal{P}(Y) \rightarrow \mathcal{P}(X)$ *such that*

$$f_{\mathcal{P}}^{-1}(B) = \{x \in X : f(x) \in B\}.$$

Again, this new function is often denoted by f^{-1}, *not to be confused with the inverse function of a bijection (which will be defined later). The set* $f^{-1}(B)$ *is called the inverse image of* B. *In particular, for* $y \in Y$, $f^{-1}(\{y\})$ *is a subset of* X.

Example 4.24. *If* $X = \{a, b, c\}$, $Y = \{1, 2, 3\}$ *and* $f(a) = 2, f(b) = f(c) = 1$, *then* $f^{-1}(\{1\}) = f^{-1}(\{1, 3\}) = \{b, c\}$, $f^{-1}(\{2\}) = f^{-1}(\{2, 3\}) = \{a\}$, $f^{-1}(\{3\}) = f^{-1}(\emptyset) = \emptyset$, $f^{-1}(\{1, 2\}) = f^{-1}(Y) = \{a, b, c\}$.

The direct image and inverse image functions have the following properties.

Theorem 4.25. *Let* $f : X \rightarrow Y$ *be a function, let* $A, B \in \mathcal{P}(X)$, *and let* $C, D \in \mathcal{P}(Y)$. *Then*
  a. $f(A \cup B) = f(A) \cup f(B)$.
  b. $f(A \cap B) \subseteq f(A) \cap f(B)$.
  c. $f^{-1}(C') = (f^{-1}(C))'$.
  d. $f^{-1}(C \cup D) = f^{-1}(C) \cup f^{-1}(D)$.
  e. $f^{-1}(C \cap D) = f^{-1}(C) \cap f^{-1}(D)$.

Proof. a. Let $y \in f(A \cup B)$. Then there is $x \in A \cup B$ such that $f(x) = y$. If $x \subset A$, then $f(x) \in f(A)$ and if $x \in B$ then $f(x) \in f(B)$. In any case $y = f(x) \in f(A) \cup f(B)$. We proved that $f(A \cup B) \subseteq f(A) \cup f(B)$.

Let $y \in f(A) \cup f(B)$. Then $y \in f(A)$ or $y \in f(B)$. There is $x \in A$ such that $f(x) = y$ or there is $x \in B$ such that $f(x) = y$. We found $x \in A \cup B$ such that $f(x) = y$, hence $y \in f(A \cup B)$. By double inclusion we get $f(A \cup B) = f(A) \cup f(B)$.

b. Let $y \in f(A \cap B)$. Then there is $x \in A \cap B$ with $f(x) = y$. It follows that $f(x) = y \in f(A) \cap f(B)$. Note that the inclusion may be strict. For $X = Y = \mathbb{R}$, $A = \{-1\}$, $B = \{1\}$ and $f(x) = x^2$ we have $f(A \cap B) = f(\emptyset) = \emptyset$, but $f(A) \cap f(B) = \{1\}$.

c. Let $x \in f^{-1}(C')$. Then $f(x) = y \in C'$, so for any $z \in C$ we have $f(x) \neq z$. This means that $x \notin f^{-1}(C)$ or $x \in (f^{-1}(C))'$.

Let $x \in (f^{-1}(C))'$. Since $x \notin f^{-1}(C)$, it follows that $f(x) \notin C$ or $f(x) \in C'$ and therefore $x \in f^{-1}(C')$. By double inclusion we get equality.

d. Let $x \in f^{-1}(C \cup D)$. Then $f(x) \in C \cup D$, so $f(x) \in C$ or $f(x) \in D$ which means that $x \in f^{-1}(C) \cup f^{-1}(D)$. For the other inclusion, let $x \in f^{-1}(C) \cup f^{-1}(D)$. It follows that $f(x) \in C$ or $f(x) \in D$, therefore $f(x) \in C \cup D$ or $x \in f^{-1}(C \cup D)$.

e. Prove as an exercise. $\qquad\square$

Example 4.26. *Let*

$$f : \mathbb{R} \to \mathbb{R}, f(x) = \begin{cases} x - 1 & \text{if } x \le 1 \\ x^2 & \text{if } x > 1 \end{cases}.$$

Find $f([-2, 3])$ *and* $f^{-1}([-3, 4])$.

Solution. We have $[-2, 3] = [-2, 1] \cup (1, 3]$ and $f([-2, 1]) = [-3, 0]$, $f((1, 3]) = (1, 9]$, hence $f([-2, 3]) = [-3, 0] \cup (1, 9]$.

Also, since $[-3, 4] = [-3, 0] \cup (0, 4]$, we have $f^{-1}([-3, 0]) = [-2, 1]$ and $f^{-1}((0, 4]) = (1, 2]$, hence $f^{-1}([-3, 4]) = [-2, 2]$. A look at the following graph is helpful.

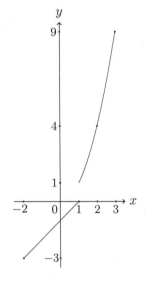

4.3   Restriction and extension of a function

Suppose $f : X \to Y$ and $g : Z \to W$ are two functions. Recall that they are equal and we write $f = g$ if and only if $X = Z$, $Y = W$ and $\forall x \in X$ we have that $f(x) = g(x)$. In particular, if we change the codomain of a function, we obtain a new function.

Example 4.27. $f : [0, \infty) \to \mathbb{R}, f(x) = \sqrt{x}$ *and* $g : [0, \infty) \to [0, \infty), g(x) = \sqrt{x}$ *are two different functions, even though they have the same domain and the same formula.*

Example 4.28. *Let* $f : \{-1, 1, 2\} \to \mathbb{R}, f(x) = x^2 - 1, g : \{-1, 1, 2\} \to \mathbb{R}, g(x) = x^3 - x^2 - x + 1$. *Then* $f(-1) = g(-1) = 0, f(1) = g(1) = 0, f(2) = g(2) = 3$, *hence* $f = g$, *even though their formulas are different. Notice that the domain has only three points. If the domain was* $\mathbb{R}$, *then they would be different functions.*

Definition 4.29. *Let* $f : X \to Y$ *be a function. A* restriction *of* f *is a function* $g : A \to Y$ *such that* $A \subseteq X$ *and* $g(a) = f(a)$ *for all* $a \in A$. *This function* g *is also denoted by* $f|_A$. *An* extension *of* f *is a function* $h : Z \to Y$ *such that* $X \subseteq Z$ *and* $h(x) = f(x)$ *for all* $x \in X$. *Sometimes an extension of* f *is denoted* $\tilde{f}$. *If* g *is a restriction of* f, *then* f *is an extension of* g.

Example 4.30. *Let* $f : \mathbb{R} \to [-1, 1], f(x) = \sin x$. *Then* $g : [-\pi/2, \pi/2] \to [-1, 1], g(x) = \sin x$ *is a restriction of* f. *This was used in trigonometry to define* $\sin^{-1} x$ *or* $\arcsin x$.
    The function $h : [-\pi, \pi] \to [-1, 1], h(x) = \sin x$ *is an extension of* g.

Example 4.31. *Consider* $f : \mathbb{R} \setminus \{0\} \to \mathbb{R}, f(x) = \dfrac{\sin x}{x}$. *We know from calculus that* $\lim\limits_{x \to 0} \dfrac{\sin x}{x} = 1$. *We can define an extension of* f *by*

$$\tilde{f} : \mathbb{R} \to \mathbb{R}, \tilde{f} = \begin{cases} \frac{\sin x}{x} & \text{for } x \neq 0 \\ 1 & \text{for } x = 0. \end{cases}$$

We say that f *was extended by continuity to* $\tilde{f}$.

We may also shrink or enlarge the codomain of a function. The new functions are sometimes called *corestriction* and *coextension*, respectively. Some people use the name restriction or extension for the new function obtained by shrinking or enlarging either the domain or codomain (or both).

Example 4.32. *Let* $f : [0, \infty) \to \mathbb{R}, f(x) = \sqrt{x}$. *Then* $g : [0, \infty) \to [0, \infty), g(x) = \sqrt{x}$ *is a corestriction of* f.

4.4 One-to-one and onto functions

Definition 4.33. *A function $f : X \to Y$ is* one-to-one *or* injective *if and only if for all $x, x' \in X$, $x \neq x' \Rightarrow f(x) \neq f(x')$.*

A function f is one-to-one if and only if for all $x, x' \in X$, $f(x) = f(x') \Rightarrow x = x'$. Indeed, this is the contrapositive of the statement $x \neq x' \Rightarrow f(x) \neq f(x')$.

Example 4.34. *The function $f : \{1, 2\} \to \{a, b, c\}, f(1) = b, f(2) = a$ is one-to-one.*

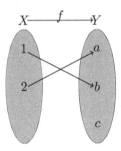

Notice that we don't have two arrows arriving at the same spot. If this happened, then f would not be one-to-one. This is because the negation of "for all $x, x' \in X$, $x \neq x' \Rightarrow f(x) \neq f(x')$" is "there exist $x, x' \in X$ with $x \neq x'$ but $f(x) = f(x')$".

Example 4.35. *The function $g : \mathbb{R} \to \mathbb{R}, g(x) = x^2$ is not one-to-one; in fact it is two-to-one for all $x \neq 0$ because $g(-x) = g(x)$.*

The restriction $g_1 : [0, \infty) \to \mathbb{R}, g_1(x) = x^2$ is one-to-one. Another one-to-one restriction is $g_2 : (-\infty, 0] \to \mathbb{R}, g_2(x) = x^2$.

Example 4.36. *As we know, $\sin : \mathbb{R} \to [-1, 1]$ is not one-to-one since for example $\sin(\pi - x) = \sin x$. But its restriction to $[-\pi/2, \pi/2]$ or to $[\pi/2, 3\pi/2]$ is one-to-one.*

Example 4.37. *Let's check if the function $f : \mathbb{R} \to \mathbb{R}, f(x) = \max(x+1, 2 - 3x)$ is one-to-one by looking at its graph.*

Solution. Since $x + 1 \geq 2 - 3x$ for $x \geq 1/4$ and $2 - 3x > x + 1$ for $x < 1/4$, we get

$$f(x) = \begin{cases} x + 1 & \text{if} \quad x \geq 1/4 \\ 2 - 3x & \text{if} \quad x < 1/4, \end{cases}$$

which has the following graph:

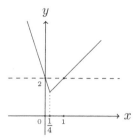

Notice that a horizontal line may cut the graph in two different points, so f is not one-to-one. Indeed, there are $x_1 \neq x_2$ such that $f(x_1) = f(x_2)$, for example $x_1 = 0, x_2 = 1$.

In general, for real functions of real variables, we have the following criterion for injectivity: any horizontal line should cut the graph in at most one point.

Definition 4.38. *A function $f : X \to Y$ is onto or* surjective *if for all $y \in Y$ there is $x \in X$ with $f(x) = y$. This is equivalent to $\mathrm{ran} f = Y$ or $f(X) = Y$.*

Example 4.39. *Let $f : X = \{1, 2, 3\}, Y = \{a, b\}, f(1) = f(2) = b, f(3) = a$ with this diagram:*

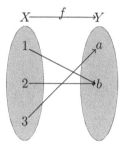

Note that at least one arrow arrives at each point in the codomain .

Example 4.40. *The function $f : \mathbb{R} \to \mathbb{R}, f(x) = x^2$ is not onto since it does not take negative values. The line $y = -1$ does not intersect the graph of f. But $g : \mathbb{R} \to [0, \infty), g(x) = x^2$ is surjective.*

Given any function $f : X \to Y$, by shrinking the codomain to $\mathrm{ran}(f)$, we can construct a surjective function $g : X \to \mathrm{ran}(f)$ such that $g(x) = f(x)$ for all $x \in X$. We distinguish between the functions f and g if f is not surjective.

We have the following criterion to check if a real function of a real variable is onto: any horizontal line through the points in the codomain should cross the graph at least once.

4.5   Composition and inverse functions

Definition 4.41. *Suppose f and g are functions. We define a new function $f \circ g$ called the* composition of f and g *with domain $\{x \in dom(g) : g(x) \in dom(f)\}$ such that $(f \circ g)(x) = f(g(x))$.*

The domain of $f \circ g$ may be the empty set, in which case $f \circ g$ is the empty function (not so interesting). To avoid this situation, many times we assume that ran(g)=dom(f). If $f, g : X \to X$, then both functions $f \circ g$ and $g \circ f$ may be defined. In general, $f \circ g \neq g \circ f$.

Example 4.42. *Let $f : [0, 4] \to [0, 2], f(x) = \sqrt{x}$ and let $g : \mathbb{R} \to [-1, \infty), g(x) = x^2 - 1$. Then*

$$f \circ g : [-\sqrt{5}, -1] \cup [1, \sqrt{5}] \to [0, 2], (f \circ g)(x) = \sqrt{x^2 - 1}$$

and

$$g \circ f : [0, 4] \to [-1, 3], (g \circ f)(x) = x - 1.$$

Example 4.43. *The function $h : (0, \pi/2) \to \mathbb{R}, h(x) = \ln(\tan x)$ is the composition $g \circ f$ where $g : (0, \infty) \to \mathbb{R}, g(y) = \ln y$ and $f : (0, \pi/2) \to (0, \infty), f(x) = \tan x$.*

Notice that $(f \circ g)(z) = \tan(\ln z)$ makes sense only for $z \in (1, e^{\pi/2})$ and $f \circ g \neq g \circ f$.

Example 4.44. *Let*

$$f : \mathbb{R} \to \mathbb{R}, f(x) = \begin{cases} x - 1 & \text{if } x \leq 1 \\ x^2 & \text{if } x > 1, \end{cases} \quad g : \mathbb{R} \to \mathbb{R}, g(x) = \begin{cases} x/2 & \text{if } x \geq 2 \\ x^3 & \text{if } x < 2. \end{cases}$$

Let us find $f \circ g$ and $g \circ f$.

Solution. We have

$$(f \circ g)(x) = \begin{cases} g(x) - 1 & \text{if } g(x) \leq 1 \\ (g(x))^2 & \text{if } g(x) > 1. \end{cases}$$

Notice that $g(x) \leq 1$ for $x \leq 1$ and for $x = 2$; otherwise $g(x) > 1$. It follows that

$$(f \circ g)(x) = \begin{cases} x^3 - 1 & \text{if } x \leq 1 \\ x^6 & \text{if } 1 < x < 2 \\ 0 & \text{if } x = 2 \\ \dfrac{x^2}{4} & \text{if } x > 2. \end{cases}$$

On the other hand,

$$(g \circ f)(x) = \begin{cases} \dfrac{f(x)}{2} & \text{if } f(x) \geq 2 \\ (f(x))^3 & \text{if } f(x) < 2. \end{cases}$$

We have $f(x) \geq 2$ for $x \geq \sqrt{2}$ and $f(x) < 2$ for $x < \sqrt{2}$, hence

$$(g \circ f)(x) = \begin{cases} \dfrac{x^2}{2} & \text{if } x \geq \sqrt{2} \\ x^6 & \text{if } 1 < x < \sqrt{2} \\ (x-1)^3 & \text{if } x \leq 1. \end{cases}$$

The operation of composition of functions preserves injectivity and surjectivity:

Theorem 4.45. *Let $f : Y \to Z$ and let $g : X \to Y$. If f and g are one-to-one functions, then so is $f \circ g : X \to Z$. If f and g are onto, then $f \circ g : X \to Z$ is also onto.*

Proof. Assume $(f \circ g)(x) = (f \circ g)(x')$, so $f(g(x)) = f(g(x'))$ for $x, x' \in X$. Since f is one-to-one, we get $g(x) = g(x')$. Since g is one-to-one, we get $x = x'$, hence $f \circ g$ is one-to-one.

Assume now that $z \in Z$. Since f is onto, we can find $y \in Y$ with $f(y) = z$. Since g is onto, there is $x \in X$ with $g(x) = y$. We conclude that $(f \circ g)(x) = z$, hence $f \circ g$ is onto. $\qquad\square$

Exercise 4.46. *Prove by counterexample that the converse of each statement of the above theorem is false.*

Definition 4.47. *A function $f : X \to Y$ is called* bijective *if it is one-to-one and onto.*

Example 4.48. *Prove that the function $f : \mathbb{R} \to \mathbb{R}, f(x) = x^3$ is bijective.*

Proof. To show that f is one-to-one, assume $x^3 = x'^3$ for some $x, x' \in \mathbb{R}$. By taking cubic roots, we get $x = x'$. To show that f is onto, let $y \in \mathbb{R}$ be arbitrary. Then $f(\sqrt[3]{y}) = (\sqrt[3]{y})^3 = y$, so we found $x = \sqrt[3]{y} \in \mathbb{R}$ with $f(x) = y$. $\qquad\square$

Definition 4.49. *We say that a function $f : X \to Y$ is* invertible *(or has an inverse) if there is $g : Y \to X$ such that $g \circ f = id_X$ and $f \circ g = id_Y$. The inverse of f is unique, is denoted f^{-1} and satisfies*

$$f^{-1}(y) = x \Leftrightarrow f(x) = y.$$

Recall that we already used the notation $f^{-1} = f_{\mathcal{P}}^{-1} : \mathcal{P}(Y) \to \mathcal{P}(X)$ for any function $f : X \to Y$ and we called it the inverse image function. If f is a bijection, we have $f^{-1}(\{y\}) = f^{-1}(y)$. You must be careful to distinguish from the context between the two meanings of f^{-1}.

We have the following characterization of invertible functions:

Theorem 4.50. *A function $f : X \to Y$ is invertible if and only if it is bijective.*

Proof. If $f : X \to Y$ is invertible, let $g : Y \to X$ be its inverse. To prove that f is one-to-one, assume $f(x_1) = f(x_2)$. Applying g to both sides, we get $(g \circ f)(x_1) = (g \circ f)(x_2)$, hence $x_1 = x_2$ since $g \circ f = id_X$. To prove that f is onto, let $y \in Y$. Then $f(g(y)) = y$, so we found $x = g(y) \in X$ such that $f(x) = y$.

Conversely, given $f : X \to Y$ bijective, define $g : Y \to X$ such that $g(y) = x \Leftrightarrow f(x) = y$. Then it is easy to verify that $g \circ f = id_X$ and $f \circ g = id_Y$, so g is the inverse of f. $\qquad\qquad\square$

Example 4.51. *The function* $\sin : [-\pi/2, \pi/2] \to [-1, 1]$ *is one-to-one and onto. Its inverse is denoted* $\sin^{-1} = \arcsin : [-1, 1] \to [-\pi/2, \pi/2]$.

Example 4.52. *Let* $f : (2, \infty) \to (-1, \infty), f(x) = \dfrac{3-x}{x-2}$. *Then f is one-to-one since* $f(x_1) = f(x_2)$ *gives*

$$\frac{3 - x_1}{x_1 - 2} = \frac{3 - x_2}{x_2 - 2},$$

which implies

$$3x_2 - x_1 x_2 - 6 + 2x_1 = 3x_1 - 6 - x_1 x_2 + 2x_2,$$

hence $x_1 = x_2$.

It is onto since given $y \in (-1, \infty)$, *the equation* $y = \dfrac{3-x}{x-2}$ *has a solution* $x = \dfrac{2y+3}{y+1}$, *which belongs to* $(2, \infty)$. *Indeed, when $y \to -1$ we have $x \to \infty$ and when $y \to \infty$, $x \to 2$. The last computation gives the formula for the inverse*

$$f^{-1} : (-1, \infty) \to (2, \infty), f^{-1}(y) = \frac{2y+3}{y+1}.$$

If we graph f and f^{-1} in the same system of coordinates, we notice that their graphs are symmetric with respect to the first quadrant bisector $y = x$.

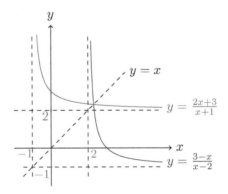

Theorem 4.53. *We have*

1. Let $f : Y \to Z, g : X \to Y$ be two functions. If f, g are bijections, then $(f \circ g)^{-1} = g^{-1} \circ f^{-1}$.

2. For an invertible function $f : X \to Y$, we have $f^{-1}(\{y\}) = \{x\}$, where $f(x) = y$.

3. If $f : X \to Y$ is invertible, then f^{-1} is the set of ordered pairs $\langle f(x), x \rangle$ in the Cartesian product $Y \times X$.

Proof. 1. Indeed, $f \circ g : X \to Z, (f \circ g)(x) = f(g(x))$ is also a bijection and $(f \circ g) \circ (g^{-1} \circ f^{-1}) = f \circ id_Y \circ f^{-1} = id_Z$ and $(g^{-1} \circ f^{-1}) \circ (f \circ g) = g^{-1} \circ id_Z \circ g = id_X$, so $(f \circ g)^{-1} = g^{-1} \circ f^{-1}$.

2. Indeed, since f is invertible, it is bijective, so given $y \in Y$ there is a unique $x \in X$ with $f(x) = y$.

3. Recall that a function $f : X \to Y$ was defined as the set of ordered pairs $\langle x, f(x) \rangle$ in the Cartesian product $X \times Y$. Now $f^{-1} : Y \to X$ has the property that $f^{-1}(y) = x$ if and only if $y = f(x)$, in other words $\langle x, f(x) \rangle \in f$ if and only if $\langle f(x), x \rangle \in f^{-1}$ since $f^{-1}(f(x)) = x$. □

Definition 4.54. *Given $f : X \to Y$, a* left inverse *or a* retract *of f is a function $r : Y \to X$ such that $r \circ f = id_X$. A* right inverse *or a* section *of f is a function $s : Y \to X$ such that $f \circ s = id_Y$.*

Example 4.55. *Let $f : \{1,2\} \to \{a,b,c\}, f(1) = b, f(2) = c$. Then $r : \{a,b,c\} \to \{1,2\}, r(a) = 1, r(b) = 1, r(c) = 2$ is a left inverse for f. Notice that $r(a)$ can be any element of $\{1,2\}$.*

Example 4.56. *Let $g : \{1,2,3\} \to \{a,b\}, g(1) = g(2) = b, g(3) = a$ and $s : \{a,b\} \to \{1,2,3\}, s(a) = 3, s(b) = 2$. Then s is a right inverse for g. Notice that to define $s(b)$, we may choose any element from the set $g^{-1}(\{b\})$.*

Theorem 4.57. *Any injective function has a left inverse. Conversely, if f has a left inverse, then f is injective.*

Proof. Given $f : X \to Y$ injective, define $r : Y \to X$ as follows. For $y \in f(X)$, say $y = f(x)$ and define $r(y) = x$. This is well defined since there is a unique x such that $f(x) = y$. For $y \in Y \setminus f(X)$, define $r(y) = x_0$ for a fixed arbitrary element of X. Then we have $r \circ f = id_X$ since $r(f(x)) = x$ for all $x \in X$. Conversely, assume f has a retract $r : Y \to X$ and let's prove that f is one-to-one. Suppose $f(x) = f(x')$. By applying r we get $r(f(x)) = r(f(x'))$ or $x = x'$, hence f is one-to-one. □

Exercise 4.58. *Let $f : \mathbb{R} \to \mathbb{R}$ be a function such that $f(2x+1) = x^2 + x - 2$ for all x. Find $f(x)$.*

Solution. Let $t = 2x + 1$. Then $x = \dfrac{t-1}{2}$ and

$$f(t) = \left(\frac{t-1}{2}\right)^2 + \frac{t-1}{2} - 2 = \frac{t^2 - 2t + 1}{4} + \frac{t-1}{2} - 2 = \frac{t^2 - 9}{4}.$$

It follows that $f(x) = \dfrac{x^2 - 9}{4}$.

4.6   *Family of sets and the axiom of choice

Definition 4.59. *Consider I, X arbitrary sets. A* family *of subsets of X is a function $f : I \to \mathcal{P}(X)$. We denote it by $\{X_i\}_{i \in I}$, where $X_i = f(i), i \in I$. The set I is called the* index set.

Many times the set X is understood, so we just talk about the family of sets $\{X_i\}_{i \in I}$. There is some ambiguity in this notation, since in an arbitrary family $\{X_i\}_{i \in I}$ we may have $i_1 \neq i_2$ and still $X_{i_1} = X_{i_2}$. This ambiguity comes from the fact that, for example, if $A \subseteq X$ and $I = \{1, 2\}$, the family with two elements $\{A_i\}_{i=1,2}$ where $A_1 = A_2 = A$, as a set is just $\{A\}$.

We define the union and the intersection of the family of sets $\{X_i\}_{i \in I}$ by

$$\bigcup_{i \in I} X_i = \{x \in X : \exists j \in I, x \in X_j\}, \quad \bigcap_{i \in I} X_i = \{x \in X : \forall j \in I, x \in X_j\}.$$

If $I = \{1, 2, 3, ..., n\}$, then we will often use the simpler notation

$$\bigcup_{i=1}^{n} X_i \quad \text{and} \quad \bigcap_{i=1}^{n} X_i.$$

If $I = \{1, 2, 3, ...\}$ is the set of positive integers, then we also use the notation

$$\bigcup_{i=1}^{\infty} X_i \quad \text{and} \quad \bigcap_{i=1}^{\infty} X_i.$$

Example 4.60. *Suppose, for each positive integer n, that $X_n = \{1, 2, 3, ..., n\}$. The index set I in this case is the set of all positive integers. Then $\displaystyle\bigcup_{n \in I} X_n = I$ and $\displaystyle\bigcap_{n \in I} X_n = \{1\}$.*

Example 4.61. *Let $I = (0, \infty)$ and let $X_i = (-i, i)$ for each $i \in I$. In this case $\displaystyle\bigcup_{i \in I} X_i = (-\infty, \infty)$ and $\displaystyle\bigcap_{i \in I} X_i = \{0\}$.*

Theorem 4.62. *We have*

1. $x \in \displaystyle\bigcup_{i \in I} X_i \Leftrightarrow \exists j \in I$ *such that $x \in X_j$.*

2. *If $\{X_i\}_{i \in I}$ and $\{X_j\}_{j \in J}$ are two family of sets, then*

$$\bigcup_{i \in I} X_i \cup \bigcup_{j \in J} X_j = \bigcup_{k \in I \cup J} X_k.$$

3. $A \cap (\bigcup_{i \in I} X_i) = \bigcup_{i \in I} (A \cap X_i)$ *for any set A.*

4. $(\bigcup_{i \in I} X_i)' = \bigcap_{i \in I} X_i'.$

Proof. 1. By definition, if $x \in \bigcup_{i \in I} X_i$, we can find $j \in I$ with $x \in X_j$. Conversely, if $x \in X_j$, then $x \in \bigcup_{i \in I} X_i$ by the definition of union.

2. Let $x \in \bigcup_{i \in I} X_i \cup \bigcup_{j \in J} X_j$. Then $x \in \bigcup_{i \in I} X_i$ or $x \in \bigcup_{j \in J} X_j$. By definition, there is $i_0 \in I$ such that $x \in X_{i_0}$ or there is $j_0 \in J$ with $x \in X_{j_0}$. In any case, there is $k_0 \in I \cup J$ (take $k_0 = i_0$ or $k_0 = j_0$) such that $x \in X_{k_0}$. We conclude that $x \in \bigcup_{k \in I \cup J} X_k$, hence $\bigcup_{i \in I} X_i \cup \bigcup_{j \in J} X_j \subseteq \bigcup_{k \in I \cup J} X_k$. The other inclusion is similar.

3. Let $x \in A \cap (\bigcup_{i \in I} X_i)$. Then $x \in A$ and $x \in \bigcup_{i \in I} X_i$, hence there is $j \in I$ with $x \in A$ and $x \in X_j$. This means $x \in A \cap X_j$, hence $x \in \bigcup_{i \in I} (A \cap X_i)$ and $A \cap (\bigcup_{i \in I} X_i) \subseteq \bigcup_{i \in I} (A \cap X_i)$ for any set A. The other inclusion is similar.

4. Let $x \in (\bigcup_{i \in I} X_i)'$. This means that for all $i \in I$ we have $x \notin X_i$, hence that for all $i \in I$ we have $x \in X_i'$. We conclude that $x \in \bigcap_{i \in I} X_i'$ and that $(\bigcup_{i \in I} X_i)' \subseteq \bigcap_{i \in I} X_i'$. Similarly, we can prove the other inclusion.

$\square$

Exercise 4.63. *a. Prove that $\bigcup_{i \in I} X_i$ is the smallest set containing all the sets X_i.*

b. Show that $\bigcap_{i \in I} X_i$ is the largest set contained in all the sets X_i.

c. State and prove the dual statements of parts 2, 3, and 4 of the above theorem.

Definition 4.64. *The Cartesian product of a family of sets $\{X_i : i \in I\}$ is*

$$\prod_{i \in I} X_i = \{x : I \to \bigcup_{i \in I} X_i : x(j) \in X_j \text{ for all } j \in I\}.$$

Exercise 4.65. *For $I = \{1, 2\}$, compare the new definition of*

$$\prod_{i=1}^{2} X_i = X_1 \times X_2$$

with the old one using ordered pairs.

We are now in the position to add a new axiom to our set theory axioms:

Axiom 3 (*Axiom of choice*). Let $\mathcal{F}$ be any family of nonempty sets. Then there is a function f defined on $\mathcal{F}$ such that $f(A) \in A$ for all $A \in \mathcal{F}$. The function f is called a choice function, and its existence may be thought of as the result of choosing, for each of the sets A in $\mathcal{F}$, an element in A.

There is, of course, no difficulty in constructing the function f if the family $\mathcal{F}$ is finite; also, if for example $\mathcal{F}$ is made of subsets of natural numbers, we can define $f(A)$ to be the minimum of A. The existence of this minimum will proved in Chapter 6. For $\mathcal{F}$ infinite we may need the axiom of choice in general. By definition, a set is finite if it is empty or it is in bijection with $\{1, 2, ..., n\}$ for some positive integer n. A set is infinite if it is not finite.

The axiom of choice can be stated as follows: For each nonempty set X, there is a function $c : \mathcal{P}(X) \setminus \{\emptyset\} \to X$ satisfying $c(A) \in A$ for every $A \in \mathcal{P}(X) \setminus \{\emptyset\}$.

Theorem 4.66. *The axiom of choice is equivalent to the fact that if none of the sets of a nonempty family $\{X_i : i \in I\}$ are empty, then the Cartesian product $\prod_{i \in I} X_i$ is not empty.*

Proof. Indeed, assuming the axiom of choice, given $\{X_i\}_{i \in I}$, we can define a function $x : I \to \bigcup_{i \in I} X_i$ such that $x(i) \in X_i$, so $\prod_{i \in I} X_i$ is not empty. Conversely, given a family $\mathcal{F}$ of nonempty sets, the fact that the Cartesian product of all members in $\mathcal{F}$ is nonempty means that we can choose, for each $A \in \mathcal{F}$, an element in A. $\square$

We will mention other equivalent statements with the axiom of choice, like Zorn's lemma and the Hausdorff maximal principle, in the next chapter. Many proofs in mathematics require the axiom of choice. For example,

Theorem 4.67. *The function $f : X \to Y$ has a right inverse iff f is surjective.*

Proof. If $f : X \to Y$ has a right inverse $s : Y \to X$, then given $y \in Y$ we can take $s(y) \in X$ such that $f(s(y)) = y$, hence f is surjective. Conversely, if $f : X \to Y$ is surjective, we can construct a section $s : Y \to X$ as follows: for each $y \in Y$ choose an element x in the nonempty set $f^{-1}(y)$, and define $s(y) = x$. This is possible by the axiom of choice. Moreover, $f(s(y)) = f(x) = y$, so s is a section for f. $\square$

Given two sets X, Y, the set of all functions from X to Y is sometimes denoted by Y^X. This notation will be useful when we will discuss cardinalities and counting techniques.

Exercise 4.68. *Find $\{a, b\}^{\{1,2,3\}}$. How many elements are in this set? How many are one-to-one? How many are onto?*

Solution. The set $\{a,b\}^{\{1,2,3\}}$ has 8 elements since we can assign to each element in the domain $\{1,2,3\}$ any of the two elements in the codomain $\{a,b\}$ and $2 \cdot 2 \cdot 2 = 8$. There are no one-to-one functions, since the values $f(1), f(2), f(3)$ cannot be distinct. There are two constant functions: $f_1(1) = f_1(2) = f_1(3) = a; f_2(1) = f_2(2) = f_2(3) = b$, which are not onto. The rest are onto.

4.7 Exercises

Exercise 4.69. *Which of the following subsets of $\{a,b,c,d\} \times \{a,b,c,d\}$ are functions? For those that fail, give a reason why.*
 1. $f = \{\langle a,b \rangle, \langle c,d \rangle\}$.
 2. $g = \{\langle a,a \rangle, \langle b,c \rangle, \langle c,a \rangle, \langle d,b \rangle\}$.
 3. $h = \{\langle a,b \rangle, \langle b,c \rangle, \langle c,d \rangle, \langle d,a \rangle, \langle a,c \rangle\}$.
 4. $k = \{\langle a,b \rangle, \langle b,c \rangle, \langle c,d \rangle, \langle d,d \rangle\}$.
 5. $\ell = \{\langle a,b \rangle, \langle b,c \rangle, \langle c,d \rangle\}$.

Exercise 4.70. *Find all functions from $\{a,b,c\}$ to $\{1,2\}$ and specify their diagrams as in Example 4.3.*

Exercise 4.71. *The equation $f(x) = \dfrac{x^2+9}{x^2-9}$ is used to define a real function of real variable. Find the largest domain of f and its graph. What is the range of f?*

Exercise 4.72. *Find $f+g, f-g, f \cdot g, f/g$ and their domains if*

$$f,g : \mathbb{R} \to \mathbb{R}, f(x) = \begin{cases} x+2 & if \quad x \le 3 \\ -x+3 & if \quad x > 3 \end{cases}, \quad g(x) = \begin{cases} x-2 & if \quad x \le 0 \\ x+1 & if \quad x > 0. \end{cases}$$

Exercise 4.73. *Prove that for $x,y \in \mathbb{R}$*

$$\max(x,y) = \frac{x+y+|x-y|}{2}, \quad \min(x,y) = \frac{x+y-|x-y|}{2}.$$

Exercise 4.74. *Let $X = \{1,3,5,6\}$, and let $Y = \{0,1,2,3,5,6\}$. Which of the following open sentences with $x \in X, y \in Y$ define a function from X to Y? How about a function from Y to X?*
 a. $x+y=6$.
 b. $y-x=1$.
 c. $y=x^2$.
 d. $x=y$.

Exercise 4.75. *Let $A = \{1,2,3,4\}$ and let $B = \{a,b,c,d,e\}$. For*

$$f : A \to B, f = \{\langle 1,e \rangle, \langle 2,c \rangle, \langle 3,a \rangle, \langle 4,e \rangle\},$$

determine the sets
  a. $f(\{1,3\})$.
  b. $f^{-1}(\{a,b,c\})$.
  c. $f^{-1}(f(\{2,4\}))$.
  d. $f(f^{-1}(\{b,d,e\}))$.

Exercise 4.76. *Let $f : \mathbb{R} \to \mathbb{R}$, $f(x) = |x - 2| + 3$. Calculate*
  a. $f((-2,5])$.
  b. $f((-2,-1) \cup (3,6))$.
  c. $f^{-1}([0,2))$.
  d. $f^{-1}([4,7))$.
  e. $f^{-1}(f((-1,3)))$.
  f. $f(f^{-1}([-1,5]))$.

Exercise 4.77. *Let $f : \{1,2\} \to \mathbb{R}$, $f(1) = 3, f(2) = 5$. Find an extension $g : \mathbb{R} \to \mathbb{R}$ of f of the form $g(x) = ax + b$.*

Exercise 4.78. *Let $f : \{-1,1,2\} \to \mathbb{R}$ such that $f(-1) = 6, f(1) = 0, f(2) = 0$. Find an extension $g : \mathbb{R} \to \mathbb{R}$ of f of the form $g(x) = ax^2 + bx + c$.*

Exercise 4.79. *Which of the following functions are one-to-one?*
  a. $g(x) = \min(-x, x), x \in \mathbb{R}$.
  b. $h(x) = \max(x + 1, 2x - 1), x \in \mathbb{R}$.
  c. $k(x) = \min(-2x, -x + 1), x \in \mathbb{R}$.

Exercise 4.80. *Let $A \subseteq \mathbb{R}$ and let $f : A \to \mathbb{R}$ as defined below. Determine two different sets A so that $f \mid_A$ is one-to-one. Choose them as large as possible.*
  a. $f(x) = \dfrac{2}{(x-3)^2}$.
  b. $f(x) = \cot x$.

Exercise 4.81. *Given three functions f, g, h, prove that $f \circ (g \circ h) = (f \circ g) \circ h$.*

Exercise 4.82. *Let $f : [1/e, e] \to [-\sin 1, \sin 1], f(x) = \sin(\ln x)$. Prove that f is a bijection and find the formula for f^{-1}.*

Exercise 4.83. *Which of the functions from $\{a,b,c,d\}$ to $\{a,b,c,d\}$ that you found in Ex. 4.69 are onto? Which are one-to-one?*

Exercise 4.84. *Find all one-to-one functions that can be defined from $\{1,2\}$ to $\{a,b,c\}$.*

Exercise 4.85. *If $f : X \to X$ is injective, prove by induction that the composition $f^n = f \circ f \circ \cdots \circ f$ is injective for all $n \geq 1$. Same for surjective.*

Exercise 4.86. *Prove that $g = \{\langle x, y \rangle \in \mathbb{R} \times \mathbb{R} : x^2 = y^2\}$ is not a function.*

Exercise 4.87. *Let $f : X \to Y$ be a function and let $A, B \in \mathcal{P}(X)$. Prove that*
  a. *If f is one-to-one, then $f(A \cap B) = f(A) \cap f(B)$.*

 b. If f is one-to-one, then $f(A') \subseteq (f(A))'$.
 c. If f is onto, then $f(A') \supseteq (f(A))'$.
 In parts b and c, find f and A such that the inclusions are strict.

Exercise 4.88. *Show that if $g \circ f$ is injective, then f is injective. Show that if $f \circ g$ is surjective, then f is surjective. Prove that the converse statements are false.*

Exercise 4.89. *Find two injective restrictions of $f : \mathbb{R} \to \mathbb{R}, f(x) = x^2 - 3x + 1$.*

Exercise 4.90. *Prove that the following functions are bijective and find their inverses.*
 1. $f : \mathbb{R} \to \mathbb{R}, f(x) = 7x + 1$.
 2. $g : (-\infty, 0] \to [0, \infty), g(x) = x^2$.
 3. $h : [2, \infty) \to (-\infty, 0], h(x) = -x^2 + 4x - 4$.
 4. $k : [-3, 1] \to [-6, 3], k(x) = \max(2x, 3x)$.
 5. $t : [0, \pi/2] \to [-1, 1], t(x) = \sin x - \cos x$.

Exercise 4.91. *Show that $f : \mathbb{R} \to \mathbb{R}, f(x) = x^2 - 6x + 2$ has invertible restrictions defined on*
 a. $(-\infty, 3]$,
 b. $[3, \infty)$, and
 c. $(-\infty, 0] \cup [3, 6)$.
 Find the inverses of these restrictions and graph them.

Exercise 4.92. *Let*

$$f : \mathbb{R} \to [0, \infty), f(x) = x^2, \quad g : \mathbb{R} \to [-1, 1], g(x) = \frac{2x}{1 + x^2},$$

$$h : [1, \infty) \to [0, \infty), h(x) = \sqrt{x - 1}.$$

Compute $f \circ g \circ h$ and specify its domain and range.

Exercise 4.93. *Let*

$$f : \mathbb{R} \to \mathbb{R}, f(x) = \begin{cases} x & \text{if } x \leq 0 \\ x + 1 & \text{if } x > 0, \end{cases} \quad g : \mathbb{R} \to [0, \infty), g(x) = \begin{cases} x^2 & \text{if } x < 0 \\ x & \text{if } x \geq 0. \end{cases}$$

Find $f \circ g$ and $g \circ f$.

Exercise 4.94. *Two functions $f, g : \mathbb{R} \to \mathbb{R}$ are such that for all $x \in \mathbb{R}$,*

$$g(x) = x^2 + x + 3, \quad (g \circ f)(x) = x^2 - 3x + 5.$$

Find all possibilities for $f(x)$.

Exercise 4.95. *Consider a family of sets $\{A_n\}_{n\geq 1}$. Prove that there is a family of disjoint sets $\{B_n\}_{n\geq 1}$ such that*

$$\bigcup_{n=1}^{\infty} A_n = \bigcup_{n=1}^{\infty} B_n.$$

Hint. Take $B_1 = A_1, B_2 = A_2 \setminus A_1, B_3 = A_3 \setminus (A_1 \cup A_2)$, etc.

Exercise 4.96. *Consider a family of sets $\{E_n\}_{n\geq 1}$ and define*

$$\limsup E_n = \bigcap_{k=1}^{\infty} \bigcup_{n=k}^{\infty} E_n, \quad \liminf E_n = \bigcup_{k=1}^{\infty} \bigcap_{n=k}^{\infty} E_n.$$

Prove that

$$\limsup E_n = \{x : x \in E_n \ \text{for infinitely many } n\},$$

$$\liminf E_n = \{x : x \in E_n \ \text{for all but finitely many } n\}.$$

Exercise 4.97. *Compute $\limsup E_n$ and $\liminf E_n$ for the following families of sets*
   a) $E_n = [1/n, n+1), n \geq 1$.
   b) $E_{2n-1} = \{-2n+1, -2n+2, ..., -1, 0, 1\}$ and $E_{2n} = \{1, 2, 3, ..., 2n\}$ for $n \geq 1$.

5

Relations

You are already familiar with certain relations like the order relation between real numbers, the relation of divisibility between integers, or the relation of inclusion between sets. Since there are other important kinds of relations, like equivalence relations, in this chapter we will define this concept in general, using its graph. We will see that a function is a special example of a relation.

After we give the formal definition of a relation, we give several examples and define operations with relations. We define equivalence relations, equivalence classes, the quotient set and the concept of partition of a set. We continue with order relations and we discuss several important elements in an ordered set. We conclude with Zermelo's well-ordering theorem, the Hausdorff maximal principle and Zorn's lemma which are equivalent to the axiom of choice.

5.1   General relations and operations with relations

Here is the formal definition of a relation.

Definition 5.1. *Consider two sets X and Y. A* relation *R from X to Y is a subset of the Cartesian product $X \times Y$. If $X = Y$, we say that R is a relation on X.*

We often write xRy (read: x is related to y) to express the fact that $\langle x, y \rangle \in R$.

Example 5.2. *Consider $X = \{1, 2, 3, 6\}$ and the relation xRy if x divides y. As a subset of $X \times X$,*

$$R = \{\langle 1, 1 \rangle, \langle 1, 2 \rangle, \langle 1, 3 \rangle, \langle 1, 6 \rangle, \langle 2, 2 \rangle, \langle 2, 6 \rangle, \langle 3, 3 \rangle, \langle 3, 6 \rangle, \langle 6, 6 \rangle\}.$$

The *domain* of R, denoted by $\mathrm{dom}(R)$, consists of all first components of pairs in R. Specifically,

$$\mathrm{dom}(R) = \{x \in X : \exists y \in Y \text{ such that } \langle x, y \rangle \in R\}.$$

In general, the domain of R may be a proper subset of X.

The range of R, written $\mathrm{ran}(R)$, is the set of all second components of pairs in R:

$$\mathrm{ran}(R) = \{y \in Y : \exists x \in X \text{ such that } \langle x, y \rangle \in R\},$$

a subset of Y.

Given a relation R from X to Y, if we let $Z = \mathrm{dom}(R) \cup \mathrm{ran}(R)$, then $R \subseteq Z \times Z$, so R can be viewed as a relation on Z.

Definition 5.3. *Consider a relation R from X to Y and subsets $A \subseteq X, B \subseteq Y$. The direct image $R(A)$ is defined as*

$$R(A) = \{y \in Y : \exists a \in A \text{ with } \langle a, y \rangle \in R\},$$

and the inverse image $R^{-1}(B)$ is defined as

$$R^{-1}(B) = \{x \in X : \exists b \in B \text{ with } \langle x, b \rangle \in R\}.$$

Of course, $R(A)$ is a subset of Y, $R^{-1}(B)$ is a subset of X, and $\mathrm{ran}(R) = R(X)$. This way, R defines a function $R : \mathcal{P}(X) \to \mathcal{P}(Y)$ and a function $R^{-1} : \mathcal{P}(Y) \to \mathcal{P}(X)$. Again we abuse the notation, since R now has two different meanings. Note that here R^{-1} denotes a set function; the inverse of a relation is defined later.

Example 5.4. *Consider $X = \{1, 2, 3, 6\}$ and the relation xRy if x divides y. Then $dom(R) = ran(R) = X$,*

$$R(\{1, 3\}) = \{1, 2, 3, 6\}, \quad R^{-1}(\{2, 3\}) = \{1, 2, 3\}.$$

Example 5.5. *Consider $X = \{1, 2, 3\}$ and $Y = \{a, b, c\}$. Then $S = \{\langle 1, b \rangle, \langle 2, a \rangle\}$ is a relation with domain $\{1, 2\}$ and range $\{a, b\}$. We have*

$$S(\{2, 3\}) = \{a\}, \quad S^{-1}(\{b, c\}) = \{1\}.$$

Example 5.6. *Any function $f : X \to Y$ is a relation from X to Y, with domain X and range a subset of Y. But not every relation is a function, because we may have xRy_1 and xRy_2 with $y_1 \neq y_2$, like in the divisibility relation from Example 5.2. The direct image and the inverse image of a function viewed as a relation coincide with those from the previous chapter.*

Definition 5.7. *Let $P(x, y)$ be a property depending on $x \in X$ and $y \in Y$. We can form*

$$\{\langle x, y \rangle \in X \times Y : P(x, y) \text{ is true}\}$$

as the set of ordered pairs satisfying the property.

Example 5.8. *Let $P(x, y)$ be $x^2 + y^2 \leq 1$ for $x, y \in \mathbb{R}$. Then*

$$T = \{\langle x, y \rangle \in \mathbb{R} \times \mathbb{R} : x^2 + y^2 \leq 1\}$$

is a relation on $\mathbb{R}$. *Note that* $dom(T) = ran(T) = [-1, 1]$. *We can visualize* T *as a disc in the plane, which could be thought as the graph of the relation.*

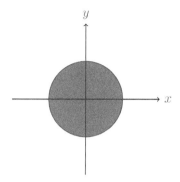

Remark 5.9. *Given a relation* $R \subseteq X \times X$ *and a set* Y *such that* $X \subset Y$, R *can be considered as a relation on* Y. *Notice that even though we may have the same set of ordered pairs, we are dealing with a different concept. For example, suppose* R *is the relation*

$$\{\langle x, y \rangle : x \leq y \text{ and } x, y \text{ are integers}\},$$

X *is the set of integers, and* Y *is the set of rational numbers. If* R *is considered as a relation on* X, *we can properly infer that for all* $x \in X$, $\langle x, x \rangle \in R$. *However if we assume that* R *is a relation on* Y, *then for example* $\langle \frac{1}{2}, \frac{1}{2} \rangle \notin R$.

Definition 5.10. *(Operations with relations). Given relations* R_i *from* X_i *to* Y_i *for* $i = 1, 2$, *their union* $R_1 \cup R_2$, *their intersection* $R_1 \cap R_2$, *and their differences* $R_1 \setminus R_2, R_2 \setminus R_1$ *are new relations as subsets of* $(X_1 \cup X_2) \times (Y_1 \cup Y_2)$. *Also, given a relation* R *from* X *to* Y, *we may consider its complement* $R' \subseteq X \times Y$, *which is another relation from* X *to* Y.

Example 5.11. *Consider* $R_1 = \{\langle u, u \rangle, \langle v, v \rangle, \langle u, v \rangle\}$ *as a relation on* $\{u, v\}$ *and* $R_2 = \{\langle v, v \rangle, \langle w, w \rangle, \langle v, w \rangle, \langle w, v \rangle\}$ *as a relation on* $\{v, w\}$. *Then*

$$R_1 \cup R_2 = \{\langle u, u \rangle, \langle v, v \rangle, \langle u, v \rangle, \langle v, v \rangle, \langle w, w \rangle, \langle v, w \rangle, \langle w, v \rangle\},$$

$$R_1 \cap R_2 = \{\langle v, v \rangle\}, \quad R_1 \setminus R_2 = \{\langle u, u \rangle, \langle u, v \rangle\}, \quad R_2 \setminus R_1 = \{\langle w, w \rangle, \langle v, w \rangle, \langle w, v \rangle\}$$

are relations on $\{u, v, w\}$. *Moreover,* $R_1' = \{\langle v, u \rangle\}$ *and* $R_2' = \emptyset$.

Definition 5.12. *The* inverse *of a relation* $R \subseteq X \times Y$ *is the relation*

$$R^{-1} = \{\langle y, x \rangle \in Y \times X : \langle x, y \rangle \in R\},$$

a subset of $Y \times X$.

Remark 5.13. *If R is a relation from X to Y, then R^{-1} is a relation from Y to X and $yR^{-1}x \Leftrightarrow xRy$. In particular, the inverse of a relation is defined always. For example, given any function $f : X \to Y$, we can form f^{-1} as a relation from Y to X. This relation is a function only in the case that f is bijective.*

Example 5.14. *For the relation $R_1 = \{\langle 1, a\rangle, \langle 2, a\rangle, \langle 1, b\rangle\}$ we have $R_1^{-1} = \{\langle a, 1\rangle, \langle a, 2\rangle, \langle b, 1\rangle\}$ and for the relation $R_2 = \{\langle v, v\rangle, \langle w, w\rangle, \langle v, w\rangle, \langle w, v\rangle\}$ we have $R_2^{-1} = R_2$.*

Example 5.15. *Let $f : \mathbb{R} \to \mathbb{R}, f(x) = x^2$. We know that f is not one-to-one, in particular, it is not an invertible function. But we can form the relation f^{-1},*

$$f^{-1} = \{\langle x^2, x\rangle : x \in \mathbb{R}\}.$$

We can visualize f and f^{-1} as subsets of $\mathbb{R}^2$:

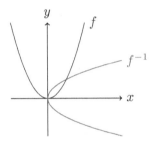

Note that the graph of f^{-1} is the reflection of the graph of f into the line $y = x$.

Theorem 5.16. *If R and S are relations from X to Y, then:*
1. *$(R^{-1})^{-1} = R$.*
2. *$(R \cup S)^{-1} = R^{-1} \cup S^{-1}$.*
3. *$(R \cap S)^{-1} = R^{-1} \cap S^{-1}$.*
4. *$(R \setminus S)^{-1} = R^{-1} \setminus S^{-1}$.*

Proof. 1. It follows from the definition of the inverse: $\langle x, y\rangle \in (R^{-1})^{-1}$ iff $\langle y, x\rangle \in R^{-1}$ iff $\langle x, y\rangle \in R$.

For 2, 3, and 4 recall that $R \cup S, R \cap S, R \setminus S$ are also relations from X to Y and by taking the inverses we just reverse the ordered pairs. □

Theorem 5.17. *If R and S are relations, then:*
1. *$dom(R \cap S) \subseteq dom(R) \cap dom(S)$.*
2. *$ran(R \cap S) \subseteq ran(R) \cap ran(S)$.*
3. *$dom(R \cup S) = dom(R) \cup dom(S)$.*
4. *$ran(R \cup S) = ran(R) \cup ran(S)$.*
5. *$dom(R) = ran(R^{-1})$.*

Proof. 1. Let $x \in \text{dom}(R \cap S)$. Then there is y such that $\langle x, y \rangle \in R \cap S$. This means that $\langle x, y \rangle \in R$ and $\langle x, y \rangle \in S$, hence $x \in \text{dom}(R) \cap \text{dom}(S)$. The reversed inclusion is false. Here is a counterexample: Let $R = \{\langle a, b \rangle, \langle b, a \rangle\}, S = \{\langle a, b \rangle, \langle b, c \rangle\}$. Then $R \cap S = \{\langle a, b \rangle\}, \text{dom}(R \cap S) = \{a\}$ and $\text{dom}(R) \cap \text{dom}(S) = \{a, b\}$.

We leave the other parts as an exercise for the reader. $\qquad\square$

Definition 5.18. *If R and S are relations, the composition of R and S, denoted $R \circ S$, is the relation*

$$\{\langle x, y \rangle : \exists z \text{ such that } \langle x, z \rangle \in S \text{ and } \langle z, y \rangle \in R\}.$$

Some people prefer the notation $S \circ R$ for the above relation, but this clashes with the notation for composition of functions, introduced in the previous chapter. Indeed, assuming that f and g are functions, then $\langle x, y \rangle \in f \circ g$ if there is $\langle x, z \rangle \in g$ and $\langle z, y \rangle \in f$, hence we have $z = g(x)$ and $y = f(g(x))$.

Notice that $\text{dom}(R \circ S) \subseteq \text{dom}(S)$ and $\text{ran}(R \circ S) \subseteq \text{ran}(R)$. In general, $R \circ S \neq S \circ R$.

Example 5.19. *With $R_1 = \{\langle 1, a \rangle, \langle 2, a \rangle, \langle 1, b \rangle\}$ and $R_2 = \{\langle x, y \rangle \in \mathbb{R} \times \mathbb{R} : x^2 + y^2 = 1\}$, $R_1 \circ R_1 = \emptyset$, $R_2 \circ R_2 = R_2$. In particular, notice that the composition of two relations may be empty.*

Example 5.20. *Let xPy if x is a parent of y, and xSy if x is a sister of y. Then $P \circ P = G$ where xGy if x is a grandparent of y and $P \circ S = A$, where xAy if x is an aunt of y.*

Theorem 5.21. *If R and S are relations, then:*
  1. $(R \circ S) \circ T = R \circ (S \circ T)$.
  2. $(R \circ S)^{-1} = S^{-1} \circ R^{-1}$.
  3. $R \circ (S \cup T) = (R \circ S) \cup (R \circ T)$.
  4. $R \circ (S \cap T) \subseteq (R \circ S) \cap (R \circ T)$.

Proof. 1. Let $\langle x, y \rangle \in (R \circ S) \circ T$. Then there is z such that $\langle x, z \rangle \in T$ and $\langle z, y \rangle \in R \circ S$. We can find $w \in \text{dom}(R) \cap \text{ran}(S)$ such that $\langle z, w \rangle \in S$ and $\langle w, y \rangle \in R$. It follows that $\langle x, w \rangle \in S \circ T$, and we conclude that $\langle x, y \rangle \in R \circ (S \circ T)$. The other inclusion is proved similarly.

2. Let $\langle x, y \rangle \in (R \circ S)^{-1}$. Then $\langle y, x \rangle \in R \circ S$, so there is z with $\langle y, z \rangle \in S$ and $\langle z, x \rangle \in R$. But then $\langle x, z \rangle \in R^{-1}$ and $\langle z, y \rangle \in S^{-1}$, so $\langle x, y \rangle \in S^{-1} \circ R^{-1}$. The other inclusion is similar.

3. The proof is by double inclusion. Let $\langle x, y \rangle \in R \circ (S \cup T)$. We can find z such that $\langle x, z \rangle \in S \cup T$ and $\langle z, y \rangle \in R$. It follows that $\langle x, y \rangle \in R \circ S$ or $\langle x, y \rangle \in R \circ T$, hence $\langle x, y \rangle \in (R \circ S) \cup (R \circ T)$. For the other inclusion, let $\langle x, y \rangle \in (R \circ S) \cup (R \circ T)$. It follows that $\langle x, y \rangle \in R \circ S$ or $\langle x, y \rangle \in R \circ T$. We can find z such that $\langle x, z \rangle \in S$ and $\langle z, y \rangle \in R$ or we can find w such that

$\langle x, w \rangle \in T$ and $\langle w, y \rangle \in R$. In each case we get $\langle x, y \rangle \in R \circ (S \cup T)$, since $\langle x, z \rangle, \langle x, w \rangle \in S \cup T$.

4. Let $\langle x, y \rangle \in R \circ (S \cap T)$. We can find z with $\langle x, z \rangle \in S \cap T$ and $\langle z, y \rangle \in R$. It follows that $\langle x, y \rangle \in R \circ S$ and $\langle x, y \rangle \in R \circ T$, hence $\langle x, y \rangle \in (R \circ S) \cap (R \circ T)$. Note that the reversed inclusion may be false.

$\square$

5.2   Equivalence relations and equivalence classes

Definition 5.22. *Suppose R is a relation on X. By definition,*

1. R is reflexive *iff $\forall x \in X$ we have xRx.*

2. R is symmetric *iff $\forall x, y \in X$ we have $xRy \Rightarrow yRx$.*

3. R is transitive *iff $\forall x, y, z \in X$ we have $xRy \wedge yRz \Rightarrow xRz$.*

4. R is an equivalence relation *if and only if R is reflexive, symmetric, and transitive.*

Sometimes an equivalence relation is denoted by $\sim$.

Example 5.23. *Let $X = \{a, b, c\}$ and let*

$$R = \{\langle a, a \rangle, \langle a, b \rangle, \langle b, a \rangle, \langle b, b \rangle, \langle c, c \rangle\}.$$

Then R is an equivalence relation on X. Indeed, it is reflexive since $\langle a, a \rangle, \langle b, b \rangle, \langle c, c \rangle \in R$. It is symmetric since $\langle a, b \rangle \in R$ implies $\langle b, a \rangle \in R$. For transitivity, we check that

$$\langle a, a \rangle, \langle a, b \rangle \in R \Rightarrow \langle a, b \rangle \in R, \langle a, b \rangle, \langle b, a \rangle \in R \Rightarrow \langle a, a \rangle \in R,$$

$$\langle a, b \rangle, \langle b, b \rangle \in R \Rightarrow \langle a, b \rangle \in R, \langle b, a \rangle, \langle a, a \rangle \in R \Rightarrow \langle b, a \rangle \in R,$$

$$\langle b, a \rangle, \langle a, b \rangle \in R \Rightarrow \langle b, b \rangle \in R, \langle b, b \rangle, \langle b, a \rangle \in R \Rightarrow \langle b, a \rangle \in R.$$

Example 5.24. *Let X be a nonempty set, and consider the equality relation on X: xRy iff $x = y$. Then R is obviously reflexive, symmetric and transitive, so equality is an equivalence relation.*

Example 5.25. *Let $X = \{a, b, c\}$ and let $S = \{\langle a, a \rangle, \langle a, b \rangle, \langle b, c \rangle\}$. Then S is not reflexive, since $\langle b, b \rangle \notin S$, is not symmetric since $\langle a, b \rangle \in S$ but $\langle b, a \rangle \notin S$, and it is not transitive since $\langle a, b \rangle, \langle b, c \rangle \in S$, but $\langle a, c \rangle \notin S$.*

Example 5.26. *Let $X = \mathbb{Z} \setminus \{0\}$ and $D = \{\langle x, y \rangle \in X \times X : x \text{ divides } y\}$. Then D is reflexive and transitive, but not symmetric because $\langle 2, 4 \rangle \in D$ but $\langle 4, 2 \rangle \notin D$.*

Example 5.27. *For $X = \mathbb{Z}$ and n a positive integer, let xRy if and only if n divides $x - y$. Then R is an equivalence relation on X, called* congruence modulo n. *Another notation for this relation is $x \equiv y (mod\, n)$.*

Indeed, R is reflexive since for all $x \in X$, n divides $x - x = 0$. If n divides $x - y$, then n divides $y - x$, so R is symmetric. If n divides $x - y$ and n divides $y - z$, then n divides $x - y + y - z = x - z$, hence R is transitive.

It turns out that there is a strong connection between equivalence relations on X and partitions of X.

Definition 5.28. *Suppose X is a set and $\mathcal{P}$ is a family of subsets of X. We say that $\mathcal{P}$ is a* partition *of X if and only if:*

1. $A \in \mathcal{P} \Rightarrow A \neq \emptyset$,

2. $\mathcal{P}$ is disjointed (any two different members of $\mathcal{P}$ have empty intersection), and

3. $\displaystyle\bigcup_{A \in \mathcal{P}} A = X$.

Example 5.29. *The family of sets $\{\{1,2,3\}, \{4,5\}, \{6\}\}$ is a partition of the set $\{1,2,3,4,5,6\}$.*

Example 5.30. *The family $\{\{1,2,3\}, \{4,5,6\}, \{6\}\}$ is not a partition of the set $\{1,2,3,4,5,6\}$; in fact, it is not a partition of any set, since $\{4,5,6\} \cap \{6\} = \{6\} \neq \emptyset$.*

Theorem 5.31. *Suppose $\mathcal{P} = \{X_i\}_{i \in I}$ is a partition of X. Then there is an equivalence relation R on X such that:*

$$xRy \Leftrightarrow \exists i \in I \text{ such that } x, y \in X_i.$$

Proof. Let $x \in X$. Since $X = \displaystyle\bigcup_{i \in I} X_i$, there is an $i_0 \in I$ with $x \in X_{i_0}$. Hence xRx, and R is reflexive. Obviously R is symmetric by definition. Let $x, y, z \in X$ with xRy and yRz. There is $i \in I$ with $x, y \in X_i$ and there is $j \in I$ with $y, z \in X_j$. Since $y \in X_i \cap X_j$ and $\mathcal{P}$ is a partition, we have $i = j$ and $X_i = X_j$, so $x, z \in X_i$ and xRz, hence R is transitive. $\qquad\square$

Definition 5.32. *Suppose X is a set and R is an equivalence relation on X. The* equivalence class *of $x \in X$, denoted $[x]$, is the set*

$$[x] = \{y \in X : xRy\}.$$

An element $a \in [x]$ is called a representative *of the class of x. The set of (distinct) equivalence classes is denoted by X/R, and it is called the* quotient set.

Theorem 5.33. *Consider an equivalence relation R on X. Then the family of equivalence classes*

$$X/R = \{[x] \mid x \in X\}$$

forms a partition of X.

Proof. Notice that each $[x]$ is a nonempty subset of X, since $x \in [x]$. Moreover, we have $X = \bigcup_{x \in X} [x]$ by double inclusion.

We want to show that two equivalence classes are either equal or disjoint. Let $[x], [y]$ be equivalence classes. If xRy, then for any $x' \in [x]$ we have xRx', and by transitivity we get $x'Ry$, hence $x' \in [y]$ and $[x] \subseteq [y]$. Similarly we get the other inclusion, hence for xRy we have $[x] = [y]$.

Suppose $\langle x, y \rangle \notin R$. If there is $z \in [x] \cap [y]$, from zRx and zRy we get xRy, a contradiction. Hence $[x] \cap [y] = \emptyset$, and we are done. $\square$

Example 5.34. *Consider $R = \{\langle a, a \rangle, \langle a, b \rangle, \langle b, a \rangle, \langle b, b \rangle, \langle c, c \rangle\}$, which is an equivalence relation on $X = \{a, b, c\}$. Then $[a] = [b] = \{a, b\}, [c] = \{c\}$ and the corresponding partition of X is $\{\{a, b\}, \{c\}\}$. The quotient set X/R is $\{[a], [c]\} = \{\{a, b\}, \{c\}\}$.*

Example 5.35. *Consider R the congruence modulo n on $\mathbb{Z}$, where $n \geq 2$. Then $\mathbb{Z}/R$ has n elements, denoted $[0], [1], [2], ..., [n-1]$. The set of equivalence classes is denoted $\mathbb{Z}_n$. When we want to emphasize n we write $[x]_n$ for the congruence class of x modulo n.*

Example 5.36. *Given a function $f : X \to Y$, consider the relation R_f on X defined by*

$$aR_f b \Leftrightarrow f(a) = f(b).$$

Then R_f is an equivalence relation. Indeed, since $f(x) = f(x)$ for all $x \in X$, R_f is reflexive. For $xR_f y$ we have $f(x) = f(y)$, which is the same as $f(y) = f(x)$, therefore R_f is symmetric. Transitivity follows since $f(x) = f(y)$ and $f(y) = f(z)$ implies $f(x) = f(z)$.

In particular, let $f(x) = x^2, x \in [-1, 1]$. Let's describe the equivalence classes for R_f and let's identify the quotient set. Since $f(-x) = f(x)$, it follows that $[x] = \{-x, x\}$ and $[-1, 1]/R_f$ can be identified with $[0, 1]$.

Definition 5.37. *(The quotient map) Given an equivalence relation R on X, the onto map $\pi : X \to X/R, \pi(x) = [x]$ is called the* quotient map.

In order to define functions on X/R, usually we start with a function on X, say $f : X \to Y$, and check if f does not depend on representatives, i.e., if $x_1 R x_2$ implies $f(x_1) = f(x_2)$. If this is the case for all $x_1, x_2 \in X$, then f induces a well-defined function $\bar{f} : X/R \to Y$ such that $\bar{f} \circ \pi = f$, where $\pi : X \to X/R$ is the quotient map. If the values of f depend on representatives, we say that $\bar{f}$ is not well-defined.

Example 5.38. *For $X = \{a, b, c\}$ and $R = \{\langle a, a \rangle, \langle a, b \rangle, \langle b, a \rangle, \langle b, b \rangle, \langle c, c \rangle\}$, let $f : X \to \{1, 2, 3\}, f(a) = f(b) = 2, f(c) = 1$. Since $f(a) = f(b)$, we get a well-defined function $\bar{f} : X/R \to \{1, 2, 3\}, \bar{f}([a]) = 2, \bar{f}([c]) = 1$.*

Definition 5.39. *Let R be an equivalence relation on X. A subset $A \subseteq X$ is called* saturated *if it is the union of some equivalence classes, in other words $(a \in A \land aRb) \Rightarrow b \in A$.*

Example 5.40. *For the congruence modulo 5 on $\mathbb{Z}$, the set $A = \{5k : k \in \mathbb{Z}\}$ is saturated, since $A = [0]$, but $B = \{0, 1, 2\}$ is not. The smallest saturated set containing B is $[0] \cup [1] \cup [2]$.*

5.3 Order relations

Definition 5.41. *A relation R on X is called* antisymmetric *if*

$$\forall x, y \in X, \; xRy \land yRx \Rightarrow x = y.$$

A relation R is an order *relation if it is reflexive, antisymmetric and transitive. Sometimes, an order relation is called an* ordering.
An order relation R on X is called a total order *if for all $x, y \in X$ we have xRy or yRx. In this case we also say that x, y are* comparable *elements. If neither xRy or yRx is true, then x, y are* incomparable.
Since not every order is total, a general order relation R on X is also called a partial order. *The pair (X, R) is called a* partially ordered set *(to keep in mind that it may not be a totally ordered set) or a* p.o. set *for short.*
Many times an order relation is denoted $\preceq$ or $\leq$.

Example 5.42. *In calculus we already used the usual order $\leq$ on the set of real numbers $\mathbb{R}$. This is a total order relation on $\mathbb{R}$. Unless specified otherwise, when we talk about real numbers, $\leq$ denotes the usual order relation. Recall that the notation $\leq$ may have a different meaning in a different context.*

Example 5.43. *Let X be a set, and let $\mathcal{R}$ be the relation on the power set $\mathcal{P}(X)$ such that $A\mathcal{R}B$ iff $A \subseteq B$. Then $\mathcal{R}$ is reflexive, antisymmetric and transitive, hence a partial order. Therefore, $(\mathcal{P}(X), \subseteq)$ is a p.o. set.*
In general, $\subseteq$ is not a total order, since for $X = \{a, b, c\}$, $\{a, b\}$ and $\{a, c\}$ are incomparable: neither $\{a, b\} \subseteq \{a, c\}$ nor $\{a, c\} \subseteq \{a, b\}$ is true.

Example 5.44. *Let $X = \mathbb{R}$ and define xRy iff $x < y$. Then R is neither reflexive or symmetric, but R is transitive.*
A transitive relation which is neither reflexive or symmetric is called a strict order. *It can be shown that $\subset$ (strict inclusion) is also a strict order on $\mathcal{P}(X)$.*

Example 5.45. *Consider $X = \mathbb{R} \cup \{-\infty, \infty\}$ with the usual order relation $\leq$ on $\mathbb{R}$ and such that by definition $-\infty \leq x \leq \infty$ for all $x \in \mathbb{R}$. Then $(X, \leq)$ becomes a totally ordered set. The set X is denoted sometimes by $\overline{\mathbb{R}}$.*

Theorem 5.46. *Consider P a relation on X which is reflexive and transitive (such a relation is called a* preorder *relation). Consider R defined by $xRy \Leftrightarrow xPy$ and yPx. Then R is an equivalence relation.*

Define now the relation S on the quotient set X/R such that $[x]S[y] \Leftrightarrow \exists x' \in [x] \, \exists y' \in [y]$ such that $x'Py'$. Then S is an order relation on X/R.

Proof. It is easy to check using the definition that R is reflexive, symmetric and transitive, so R is an equivalence relation.

Let's prove now that S is an order relation. We have $[x]S[x]$ since xPx, so S is reflexive. Assume $[x]S[y]$ and $[y]S[x]$. By definition, there are $x' \in [x], y' \in [y]$ such that $x'Py'$ and there are $y'' \in [y], x'' \in [x]$ such that $y''Px''$. By definition of R, we have $y'Py'', y''Py', x'Px''$ and $x''Px'$. In particular, $y'Py'', y''Px''$, and $x''Px'$. By transitivity of P, it follows that $y'Px'$, hence since also $x'Py'$, we get $x'Ry'$ and $[x] = [y]$. This proves that S is antisymmetric. Transitivity of S is left as an exercise.

$\square$

Example 5.47. *Let $\mid$ be the divisibility relation on $\mathbb{Z} \setminus \{0\}$, i.e., $x \mid y$ if there is $z \in \mathbb{Z} \setminus \{0\}$ such that $y = xz$. Then $\mid$ is a preorder relation. The associated equivalence relation has classes $[x] = \{-x, x\}$, and $\mathbb{Z}/R$ can be identified with the set of positive integers, denoted $\mathbb{P}$. The associated partial order is divisibility on $\mathbb{P}$.*

Given an order relation R on X, we can define a new order relation by taking R^{-1}. This is called the opposite order. For example, if $\leq$ is the usual order on $\mathbb{R}$, then $\leq^{-1}$ is the same as $\geq$.

5.4 *More on ordered sets and Zorn's lemma

In this section, unless otherwise specified, $\preceq$ denotes an order relation, and $\prec$ is the corresponding strict order.

Definition 5.48. *Let $(X, \preceq)$ be a partially ordered set.*
A least element *or* minimum *is $y \in X$ such that $y \preceq x$ for all $x \in X$.*
A minimal element *is $y \in X$ such that there is no $z \in X$ with $z \prec y$.*
A greatest element *or* maximum *is $y \in X$ such that $x \preceq y$ for all $x \in X$.*
A maximal element *is $y \in X$ such that there is no $z \in X$ with $y \prec z$.*
An upper bound *for $A \subseteq X$ is $x \in X$ with $a \preceq x$ for all $a \in A$.*
A lower bound *for $A \subseteq X$ is $x \in X$ with $x \preceq a$ for all $a \in A$.*

The least upper bound *or* supremum *of a set* $A \subseteq X$ *is* $x \in X$ *such that* $a \preceq x$ *for all* $a \in A$ *and if* $a \preceq y$ *for all* $a \in A$, *then* $x \preceq y$.

The greatest lower bound *or* infimum *of a set* $A \subseteq X$ *is* $x \in X$ *such that* $x \preceq a$ *for all* $a \in A$ *and if* $y \preceq a$ *for all* $a \in A$, *then* $y \preceq x$.

The minimum and the maximum in a p.o. set $(X, \preceq)$ (if they exist) are unique. In that case the minimum is the only minimal element, and the maximum is the only maximal element. A p.o. set $(X, \preceq)$ may have several minimal and maximal elements.

One can define the minimum, the maximum, minimal and maximal elements of an arbitrary subset $A \subseteq X$. A subset $A \subseteq X$ may have several upper bounds and several lower bounds (if any).

The least upper bound of A (if it exists) is unique and is denoted lub A or sup A. The greatest lower bound is also unique (if it exists) and is denoted glb A or inf A.

Example 5.49. $(\mathcal{P}(\{1,2,3\}) \setminus \{\emptyset\}, \subseteq)$ *is a p.o. set. There is no minimum, and* $\{1\}, \{2\}, \{3\}$ *are minimal elements. The maximum is* $\{1,2,3\}$. *The upper bounds of* $A = \{\{1\}, \{2\}\}$ *are* $\{1,2\}, \{1,2,3\}$ *and the supremum of* A *is* $\{1,2\}$. *The set* A *has no lower bounds and no infimum.*

Exercise 5.50. *Let* $X = \{2,3,4,6,8,12\}$ *with the divisibility relation* $|$. *Find the minimal elements and the maximal elements in the p.o. set* $(X, |)$. *Find the lower bounds and the upper bounds of* $A = \{3,4\} \subseteq X$ *(if any). Does* $(X, |)$ *have a minimum or a maximum?*

Solution. The minimal elements are $2, 3$ because there is no $x \in X$ with $x \mid 2$ or $x \mid 3$ and $x \neq 2, x \neq 3$. The maximal elements are $8, 12$ since there is no other $y \in X$ such that $8 \mid y$ or $12 \mid y$. The set $A = \{3, 4\}$ has no lower bound since there is no $a \in X$ with $a \mid 3$ and $a \mid 4$. The only upper bound of A is 12. There is no minimum or maximum.

A good way to visualize $(X, |)$ is the following graph, in which the elements of X are vertices joined by an edge if they are related:

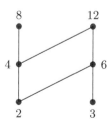

Note the position of the minimal elements at the bottom and of the maximal elements at the top.

Definition 5.51. *Let* $(X, \preceq)$ *be a p.o. set. If* $\{x, y\} \subseteq X$ *has a least upper* *bound, this element is denoted $x \vee y$ (not to be confused with the logic disjunction symbol). Similarly, $x \wedge y$ denotes the greatest lower bound of $\{x, y\}$. We say that a p.o. set $(X, \preceq)$ is a* lattice *if $x \vee y$ and $x \wedge y$ exist for all $x, y \in X$.*

Example 5.52. *The p.o. set $(\mathcal{P}(X), \subseteq)$ is a lattice with $A \vee B = A \cup B, A \wedge B = A \cap B$ for $A, B \in \mathcal{P}(X)$. The set $(\mathbb{N}, |)$ is a lattice with $a \vee b = lcm(a, b), a \wedge b = gcd(a, b)$, where lcm denotes the least common multiple, and gcd denotes the greatest common divisor (we learn more about lcm and gcd in a later chapter).*

Definition 5.53. *Consider $(X, \preceq)$ and $(X', \preceq')$ two p.o. sets. A function $f : X \to X'$ is called* increasing *if $x_1 \preceq x_1$ implies $f(x_1) \preceq' f(x_2)$ for all $x_1, x_2 \in X$. An* isomorphism *of p.o. sets is a function $f : X \to X'$ which has an increasing inverse $f^{-1} : X' \to X$.*

Example 5.54. *The function $f : \mathbb{R} \to \mathbb{R}, f(x) = x^3$ is increasing. In fact f is an isomorphism of ordered sets, since its inverse $f^{-1} : \mathbb{R} \to \mathbb{R}, f^{-1}(y) = \sqrt[3]{y}$ is also increasing. The function $g : \mathbb{R} \to \mathbb{R}, g(x) = x^2$ is not increasing, but it has an increasing restriction $g_1 : [0, \infty) \to [0, \infty), g_1(x) = x^2$.*

Example 5.55. *Consider the p.o. sets $(\mathcal{P}(\{a, b\}), \subseteq)$ and $(\{1, 2, 3, 6\}, |)$. Then $f : \mathcal{P}(\{a, b\}) \to \{1, 2, 3, 6\}, f(\emptyset) = 1, f(\{a\}) = 2, f(\{b\}) = 3, f(\{a, b\}) = 6$ is an isomorphism of p.o. sets.*

Example 5.56. *We have seen that $\bar{\mathbb{R}} = \mathbb{R} \cup \{-\infty, \infty\}$ is a totally ordered set, where $\mathbb{R}$ has the usual order and $-\infty < x < \infty$ for all $x \in \mathbb{R}$. In $\bar{\mathbb{R}}$, every subset has an upper bound and a lower bound. Indeed, $-\infty$ is a minimum, and ∞ is a maximum. As a consequence, the totally ordered sets $\mathbb{R}$ and $\bar{\mathbb{R}}$ are not isomorphic, since for example $[0, \infty)$ has no upper bound in $\mathbb{R}$.*

Definition 5.57. *We say that $(X, \preceq)$ is* well-ordered *if every nonempty subset A of X has a minimum (a lower bound for A belonging to A). This element is also called the* first element *of A or the smallest element of A.*

It is easy to see that a well-ordered set $(X, \preceq)$ is totally ordered. Indeed, given $x, y \in X$, the set $\{x, y\}$ has a smallest element, say x. Then $x \preceq y$.

Example 5.58. *Any subset of $\mathbb{N}$ (including $\mathbb{N}$ itself) with the natural order is well-ordered. We will prove this in the chapter about positive integers, using the Peano axioms.*

Example 5.59. *The sets $\mathbb{Z}$ and $\mathbb{R}$ with the usual order are totally ordered, but not well-ordered, since for example $\{..., -3, -2, -1, 0\}$ has no smallest element. Also, $\bar{\mathbb{R}} = \mathbb{R} \cup \{-\infty, \infty\}$ is not well-ordered.*

Example 5.60. *Consider $X = \mathbb{N} \cup \{\omega\}$, where $\omega \notin \mathbb{N}$. Consider the natural order on $\mathbb{N}$ and let $n < \omega$ for all $n \in \mathbb{N}$. Then X becomes a well-ordered set, not isomorphic to $\mathbb{N}$, since X has a maximum.*

Theorem 5.61. *(Principle of induction for well-ordered sets) Let $(X, \preceq)$ be well-ordered and let $A \subseteq X$ such that for all $x \in X$, whenever $a \prec x$ for all $a \in A$, we have $x \in A$. Then $A = X$.*

Proof. Assume $A \neq X$ and let $B = X \setminus A \neq \emptyset$. Then B has a smallest element, call it b_0. It follows that $a \prec b_0$ for all $a \in A$, so by hypothesis $b_0 \in A$, a contradiction. □

We state without proof

Theorem 5.62. *(Zermelo's well-ordering theorem). Every nonempty set X has an order $\preceq$ such that $(X, \preceq)$ is well-ordered.*

This seems very reasonable for finite sets and even for some infinite sets (we already mentioned that $\mathbb{N}$ is well-ordered), but for sets like the interval $(1, \infty)$, it seems entirely unreasonable, and can be proved only using the axiom of choice. We know that with the usual order, there is no smallest element in $(1, \infty)$.

Definition 5.63. *A chain in a p.o. set $(X, \preceq)$ is a totally ordered subset of X.*

Theorem 5.64. *(The Hausdorff maximal principle) Every p.o. set $(X, \preceq)$ has a maximal chain (with respect to inclusion).*

A consequence of the Hausdorff maximal principle is

Theorem 5.65. *(Zorn's lemma). Let $(X, \preceq)$ be a partially ordered set such that every chain has an upper bound. Then X has a maximal element.*

Proof. Indeed, an upper bound for a maximal chain of X is a maximal element of X. □

Example 5.66. *Consider X a nonempty set. Then $(\mathcal{P}(X), \subseteq)$ has the property that every chain has an upper bound. Indeed, given a totally ordered family $\{C_i\}_{i \in I}$ of $\mathcal{P}(X)$, we can take the union $C = \bigcup_{i \in I} C_i$, which is an upper bound for $\{C_i\}_{i \in I}$.*

In fact, the last two results are equivalent. Applying Zorn's lemma to the collection of totally ordered subsets of X, which is partially ordered by inclusion, we can prove the Hausdorff maximal principle.

Zorn's lemma is a very important tool in various parts of mathematics. For example, it is used to prove the existence of maximal ideals in a commutative ring with identity, the existence of a basis in an arbitrary vector space and the existence of a spanning tree in a graph. We have the following result.

Theorem 5.67. *The following are equivalent:*

(i) The axiom of choice (in the form: if $\{X_i\}_{i \in I}$ is a nonempty collection of nonempty sets, then $\prod_{i \in I} X_i$ is nonempty).

(ii) Zermelo's well-ordering theorem.

(iii) The Hausdorff maximal principle.

(iv) Zorn's lemma.

Proof. (partial) We have seen already that (iii) and (iv) are equivalent. We prove now the implication $(iv) \Rightarrow (ii)$. Let $\mathcal{W}$ be the collection of well orders on subsets of X, and define a partial order $\preceq$ on $\mathcal{W}$ as follows. If R_1 and R_2 are well orders on the subsets $E_1, E_2 \subseteq X$, then $R_1 \preceq R_2$ if $E_1 \subseteq E_2$, R_2 restricted to E_1 agrees with R_1, and if $y \in E_2 \setminus E_1$, then $x R_2 y$ for all $x \in E_1$. It is easy to see that $\mathcal{W}$ is not empty and that $(\mathcal{W}, \preceq)$ satisfies the hypotheses of Zorn's lemma. Indeed, if $\mathcal{T} \subseteq \mathcal{W}$ is totally ordered, by taking the union of all sets in $\mathcal{T}$ with the appropriate well order, we get an upper bound for $\mathcal{T}$. We deduce that $\mathcal{W}$ has a maximal element (E, R). We must have $E = X$, since if $x_0 \in X \setminus E$, then R can be extended to a well order $\tilde{R}$ on $E \cup \{x_0\}$ by taking $\tilde{R} = R$ on E and $x \tilde{R} x_0$ for all $x \in E$.

For $(ii) \Rightarrow (i)$, let $X = \bigcup_{i \in I} X_i$ and choose a well ordering on X. For $i \in I$, let $f : I \to X$ such that $f(i)$ is the minimal element of X_i. Then $f \in \prod_{i \in I} X_i$.

$\square$

We have now in our toolbox these equivalent statements from the previous theorem. Not all mathematicians assume the axiom of choice, but this would limit drastically our horizons. For example, without the axiom of choice, one cannot prove the existence of a set of real numbers that is not Lebesgue measurable. Also, the Banach–Tarski paradox uses the axiom of choice. This paradox says that one can decompose a three-dimensional solid ball into finitely many pieces, which can be reassembled into two solid balls each with the same volume as the original ball.

Definition 5.68. *A directed set (or filtered set) is a preordered set $(X, \preceq)$ (recall that $\preceq$ is reflexive and transitive) such that for any $a, b \in X$ there is $c \in X$ with $a \preceq c$ and $b \preceq c$.*

Example 5.69. *The ordered set $(\mathbb{N}, |)$ is directed, since given $a, b \in \mathbb{N}$ we can take $c = lcm(a, b)$. Let $X = \{-2, 2, 3, 4\}$. Then the preordered set $(X, |)$ is not directed, since the set $\{-2, 3\}$ has no upper bound in X.*

Example 5.70. *Given a set X, $(\mathcal{P}(X), \subseteq)$ is a directed set, since given $A, B \in \mathcal{P}(X)$, we can take $C = A \cup B$.*

Theorem 5.71. *In a directed set $(X, \preceq)$, any finite subset with $n \geq 2$ elements has an upper bound.*

Proof. Let $A \subseteq X$ with $n \geq 2$ elements. We use induction on n. For $n = 2$, we have an upper bound by definition. Assume that this is true for subsets with k elements, and let us prove it for $\{x_1, x_2, ..., x_k, x_{k+1}\}$. Consider y an upper bound for $\{x_1, x_2, ..., x_k\}$. By definition, the set $\{y, x_{k+1}\}$ has un upper bound z. By transitivity, we get that z is an upper bound for $\{x_1, x_2, ..., x_k, x_{k+1}\}$. $\square$

Definition 5.72. *A* net *(or generalized sequence) in a set Y is a function $x : (I, \preceq) \to Y$, where $(I, \preceq)$ is a directed set. If $I = \mathbb{N}$ with the usual order, then x is called a sequence in Y. We denote a net by $(x_i)_{i \in I}$, where $x_i = x(i)$ for all $i \in I$. When $I = \mathbb{N}$, we also write $(x_n)_{n \geq 0}$ or just (x_n).*

Example 5.73. *Let $I = \mathbb{R}$ with the usual order and for $i \in I$ let $x_i = [i-1, i)$. Then $(x_i)_{i \in I}$ is a generalized sequence in $\mathcal{P}(\mathbb{R})$.*

Example 5.74. *A constant sequence is $x : I \to Y$ such that $x_i = y_0$ for all $i \in I$ for some fixed element $y_0 \in Y$. Note that we distinguish the constant sequence $(y_0)_{i \in I}$ from the set $\{y_0\}$.*

Generalized sequences are used in general topology and in functional analysis.

5.5 Exercises

Exercise 5.75. *Let $A = \{2, 3, 4\}$ and let $B = \{2, 6, 12, 17\}$. List the elements of*
$$R = \{\langle x, y \rangle \in A \times B : x \mid y\}.$$
Find R^{-1}, $\mathrm{dom}(R)$ and $\mathrm{ran}(R)$.

Exercise 5.76. *Let $A = \{1, 2, 3, 4\}$ and consider*
$$R = \{\langle x, y \rangle \in A \times A : y - x \text{ is even}\}.$$
Find the elements of R and find $R \circ R$.

Exercise 5.77. *Let S be the relation from $\{a, b, c, 2, 3\}$ to $\{a, b, e, f, 3, 6\}$ given by*
$$S = \{\langle c, a \rangle, \langle b, 3 \rangle, \langle 3, e \rangle, \langle 2, b \rangle, \langle a, f \rangle, \langle b, 6 \rangle\}.$$
Find $\mathrm{dom}(S)$, $\mathrm{ran}(S)$ and S^{-1}.

Exercise 5.78. *Find the domain and the range for the following relations on $\mathbb{R}$ and then graph them:*
a. $R_1 = \{\langle x, y \rangle \in \mathbb{R} \times \mathbb{R} : x = -2y^2 + 3\}$.
b. $R_2 = \{\langle x, y \rangle \in \mathbb{R} \times \mathbb{R} : x = \sqrt{1 - y^2}\}$.
c. $R_3 = \{\langle x, y \rangle \in \mathbb{R} \times \mathbb{R} : x = -3 \vee |y| < 4\}$.

    d. $R_4 = \{\langle x, y \rangle \in \mathbb{R} \times \mathbb{R} : x = -3 \wedge |y| < 4\}$.

    e. $R_5 = \{\langle x, y \rangle \in \mathbb{R} \times \mathbb{R} : x^2 + y^2 < 9\}$.

    f. $R_6 = \{\langle x, y \rangle \in \mathbb{R} \times \mathbb{R} : |x| + |y| \leq 9\}$.

Exercise 5.79. *Find the inverses of the following relations:*

    a. $R = \{\langle a, 1 \rangle, \langle 2, b \rangle, \langle 3, 4 \rangle, \langle x, y \rangle\}$.

    b. $S = \{\langle x, y \rangle \in \mathbb{Z} \times \mathbb{Z} : x^2 + y^2 = 1\}$.

    c. $T = \{\langle x, y \rangle \in \mathbb{R} \times \mathbb{R} : 3x^2 - 4y^2 = 9\}$.

    d. $V = \{\langle x, y \rangle \in \mathbb{R} \times \mathbb{R} : y < 2x - 5\}$.

    e. $W = \{\langle x, y \rangle \in \mathbb{R} \times \mathbb{R} : y(x + 3) = x\}$.

Exercise 5.80. *Let* $R = \{\langle 1, 4 \rangle, \langle 2, 3 \rangle, \langle 5, 4 \rangle, \langle 3, 2 \rangle\}$, *let* $S = \{\langle 5, 1 \rangle, \langle 2, 4 \rangle, \langle 3, 3 \rangle\}$, *and let* $T = \{\langle 1, 2 \rangle, \langle 2, 1 \rangle\}$. *Compute* $R \circ S, S \circ T, R \circ R, T \circ T, R \circ (S \circ T), (R \circ S) \circ T, (R \circ S)^{-1}, R^{-1} \circ S^{-1}$.

Exercise 5.81. *Use set-builder notation to describe* $S \circ R$ *for the given relations*

    a. $R = \{\langle x, y \rangle \in \mathbb{R} \times \mathbb{R} : y = 2x - 1\}, S = \{\langle x, y \rangle \in \mathbb{R} \times \mathbb{R} : 2x^2 + 3y^2 = 5\}$

    b. $R = \{\langle x, y \rangle \in \mathbb{R} \times [0, \infty) : x = \sqrt{y}\}, S = \{\langle x, y \rangle \in \mathbb{R} \times \mathbb{R} : y = \sin x\}$.

Exercise 5.82. *Let* P *be the set of all living people, and consider the relations*

$$B = \{\langle x, y \rangle \in P \times P : y \text{ is a brother of } x\},$$

$$F = \{\langle x, y \rangle \in P \times P : y \text{ is the father of } x\},$$

$$M = \{\langle x, y \rangle \in P \times P : y \text{ is the mother of } x\},$$

and

$$S = \{\langle x, y \rangle \in P \times P : y \text{ is a sister of } x\}.$$

Describe the relations $F \circ F, M \circ F, F \circ M, M \circ B, B \circ M, F \circ S, S \circ M, M \circ S$.

Exercise 5.83. *Consider* R *a relation on* X. *Fill in the blank and prove the resulting statement:*

    a) *We have* $R \circ R \subseteq R$ *iff* _____.

    b) R *is symmetric iff the complement* R' _____.

Exercise 5.84. *Which of the following relations on* $\mathbb{R}$ *are reflexive? Which are symmetric? Which are transitive?*

    a) xRy *iff* $x - y \in \mathbb{Q}$.

    b) xRy *iff* $x - y \in \mathbb{R} \setminus \mathbb{Q}$.

    c) xRy *iff* $|x - y| \leq 2$.

Exercise 5.85. *On a set of your choice, give examples of relations that possess exactly one or exactly two of these properties: reflexivity, symmetry, transitivity.*

Exercise 5.86. *Let U be a nonempty set. Describe the properties (reflexivity, symmetry, transitivity) of the following relations on $\mathcal{P}(U)$:*
  a) $A\mathcal{R}B$ iff $A \cap B = \emptyset$.
  b) $A\mathcal{R}B$ iff $A \cap B \neq \emptyset$.
  c) $A\mathcal{R}B$ iff $A \Delta B = \emptyset$.
  d) $A\mathcal{R}B$ iff $A \setminus B$ is finite.
  e) $A\mathcal{R}B$ iff $A \Delta B$ is finite.

Exercise 5.87. *Let R and S be relations on X. Prove or disprove:*
  a) If R and S are reflexive, then $R \cap S, R \cup S$ are reflexive.
  b) If R and S are symmetric, then $R \cap S, R \cup S$ are symmetric.
  c) If R and S are antisymmetric, then $R \cap S, R \cup S$ are antisymmetric.
  d) If R and S are transitive, then $R \cap S, R \cup S$ are transitive.

Exercise 5.88. *Find two relations on $\{1, 2, 3\}$ which are not reflexive, but their composition is reflexive.*

Exercise 5.89. *Let $S = \{1, 2, 3, 4\}$, and suppose that $\sim$ is an equivalence relation on S. You know that $1 \sim 2$ and $2 \sim 3$. Describe all possibilities for $\sim$.*

Exercise 5.90. *Let X be a nonempty set and let R be an equivalence relation on X which is also a function on X. Describe the relation R.*

Exercise 5.91. *For $x, y \in \mathbb{Z}$ define xRy if 4 divides $x + 3y$. Prove that R is an equivalence relation and describe its quotient set $\mathbb{Z}/R$.*

Exercise 5.92. *For $g : [0, 2\pi] \to [-1, 1], g(x) = \sin x$, determine the relation $R_g = \{\langle x, y \rangle \in [0, 2\pi] \times [0, 2\pi] : g(x) = g(y)\}$ and $[0, 2\pi]/R_g$.*

Exercise 5.93. *Prove that $f : \mathbb{Z}_4 \to \mathbb{Z}_2, f([x]_4) = [x]_2$ is a well-defined function.*

Exercise 5.94. *Prove that $f : \mathbb{Z}_3 \to \mathbb{Z}_5, f([x]_3) = [x]_5$ is not a well-defined function.*

Exercise 5.95. *Let $f : \mathbb{Z}_5 \to \mathbb{Z}_5, f([x]) = [2x+3]$. Prove that f is well defined and determine whether f is one-to-one and onto.*

Exercise 5.96. *Given an equivalence relation on X and $B \subseteq X$ an arbitrary subset, prove that there is $A \subseteq X$ saturated such that $B \subseteq A$.*

Exercise 5.97. *Let D be the set of differentiable functions on $\mathbb{R}$. Define $f \sim g$ iff $f' = g'$. Prove that $\sim$ is an equivalence relation and describe $[f] \in D/\sim$.*

Exercise 5.98. *Construct all possible quotient sets of $\{1, 2, 3\}$*

Exercise 5.99. *Let $\mathcal{T}$ be the set of triangles in the plane and define the relation $\sim$ on $\mathcal{T}$ by $t_1 \sim t_2$ if t_1, t_2 are similar. Prove that $\sim$ is an equivalence relation and describe $[t] \in \mathcal{T}/\sim$.*

Exercise 5.100. *Let X be the unit circle $\{\langle x, y\rangle \in \mathbb{R}^2 : x^2 + y^2 = 1\}$ and define R on X by $\langle x_1, y_1\rangle R\langle x_2, y_2\rangle$ if $\langle x_1, y_1\rangle = \pm\langle x_2, y_2\rangle$. Prove that R is an equivalence relation and describe X/R.*

Exercise 5.101. *Given an order relation R on X, define a new relation S on X such that xSy if xRy and $x \neq y$. Prove that the relation S is transitive. The relation S is called the strict order associated to R.*

Exercise 5.102. *Let R and S be partial orders on X. Prove or disprove:*
  a) $R \circ S$ is a partial order on X.
  b) $R \cup S$ is a partial order on X.
  c) $R \cap S$ is a partial order on X.

Exercise 5.103. *On $\mathbb{R}^2$ we define the relation R such that $\langle x, y\rangle R\langle x', y'\rangle$ if $x \leq x'$ or $y \leq y'$. Is R an order relation?*

Exercise 5.104. *How many total order relations can be defined on $\{a, b, c\}$?*

Exercise 5.105. *Let $(X, \leq)$ be a p.o. set and let Y be an arbitrary set. On X^Y (the set of all functions $f : Y \to X$) we define $\preceq$ by $f_1 \preceq f_2$ iff $f_1(y) \leq f_2(y)$ for all $y \in Y$. Prove that $(X^Y, \preceq)$ is a p.o. set.*

Exercise 5.106. *Let $\mathbb{N} = \{0, 1, 2, 3, ...\}$ be the set of natural numbers. Consider the relation $\preceq$ on $\mathbb{N}^2$, where $(x, y) \preceq (x', y')$ iff $x \leq x'$ and $y \leq y'$. Here $\leq$ denotes the usual order. Prove that $\preceq$ is a partial order. Find the minimal elements. Find a set $A \subseteq \mathbb{N}^2$ such that any two elements of A are incomparable.*

Exercise 5.107. *Show that $X = \{1, 2, 3, 4, 6, 8, 12, 24\}$ with the divisibility relation is a lattice. How about $(Y, |)$ where $Y = \{2, 3, 4\}$?*

Exercise 5.108. *Prove that the sets $\mathbb{Z} \times \mathbb{Z}$ and $\mathbb{R} \times \mathbb{R}$ with product order*

$$\langle x_1, x_2\rangle \preceq \langle y_1, y_2\rangle \text{ whenever } x_1 \leq y_1 \text{ and } x_2 \leq y_2$$

are lattices, but $\mathbb{Z} \times \mathbb{Z}$ and $\mathbb{R} \times \mathbb{R}$ are not totally ordered. If instead we use the lexicographic or dictionary order,

$$\langle x_1, x_2\rangle \preceq_L \langle y_1, y_2\rangle \text{ if either } x_1 < y_1 \text{ or } x_1 = y_1 \text{ and } x_2 \leq y_2,$$

prove that $\mathbb{Z} \times \mathbb{Z}$ and $\mathbb{R} \times \mathbb{R}$ become totally ordered.

Exercise 5.109. *Prove that $\mathbb{N} \times \mathbb{N}$ with the lexicographic order defined above is well-ordered.*

Exercise 5.110. *Prove that the sets $\mathbb{Z} \times \mathbb{Z}$ and $\mathbb{R} \times \mathbb{R}$ with the lexicographic order are not well-ordered.*

6

Axiomatic theory of positive integers

The positive integers $1, 2, 3, ...$, sometimes called the counting numbers or the (nonzero) natural numbers, are undoubtedly the oldest numbers known to us. They are the first numbers we learn about in elementary school and their properties and the ways in which we calculate with them are among the most familiar of mathematical notions. Even so, if pressed to actually define them, most of us would find it difficult to come up with an adequate description without resorting to hand-waving or merely giving examples.

We will say enough about the positive integers through the axioms such that all their basic properties can be proved as theorems. In fact, as you will see, even though the axioms involve no explicit operation, we will be able to define and prove theorems about addition, subtraction, multiplication, division, etc. We will define the natural order relation and prove the well-ordering principle. In the process, you will learn more about mathematical induction and some useful techniques in theorem proving by using the axioms for positive integers.

Later, we will relate the positive integers with the notion of cardinality, a concept defined in a separate chapter. To get a flavor of this, we will associate 1 with the set $\{\emptyset\}$, 2 with the set $\{\emptyset, \{\emptyset\}\}$, 3 with the set $\{\emptyset, \{\emptyset, \{\emptyset\}\}\}$, and in general $n + 1$ with the set $n \cup \{n\}$. The idea will be that n represents all sets containing exactly n elements.

6.1 Peano axioms and addition

We denote here the set of positive integers by $\mathbb{P}$. Some people prefer the notation $\mathbb{N}$, but we will reserve this for $\mathbb{P} \cup \{0\}$. The set of positive integers satisfy the following five axioms, known as the Peano axioms.

Axiom 1. There exists $1 \in \mathbb{P}$.

Axiom 2. For each $n \in \mathbb{P}$, there is an element $s(n) \in \mathbb{P}$, called the successor of n.

Axiom 3. For each $n \in \mathbb{P}$, $s(n) \neq 1$.

Axiom 4. If $m, n \in \mathbb{P}$ and $s(m) = s(n)$, then $m = n$.

Axiom 5. If Q is a subset of $\mathbb{P}$ such that $1 \in Q$ and $s(n) \in Q$ whenever $n \in Q$, then $Q = \mathbb{P}$.

The first axiom implies that $\mathbb{P}$ is not empty. Axioms 2, 3, and 4 say that there is a function $s : \mathbb{P} \to \mathbb{P}$ which is one-to-one and 1 is not in the range of s. In fact 1 is the only element not belonging to the range of s.

We can define 2 as $s(1)$, 3 as $s(2)$, and in general $n + 1$ as $s(n)$. We get that $\mathbb{P} = \{1, s(1), s(s(1)), ...\} = \{1, 2, 3, ...\}$. Indeed, $\mathbb{P}$ has no other elements by Axiom 5. Later we will add the number 0 to $\mathbb{P}$ to get the set of natural numbers, denoted by $\mathbb{N}$.

Axiom 5 is also called the *principle of mathematical induction*. As we already mentioned in the chapter about proofs, it gives us a method to prove many statements of the form $\forall n \ S(n)$. The basic idea behind its use is as follows. First, it makes sense to discuss the set of all values of n for which the open sentence $S(n)$ is true. This "truth set" may consist of all, some, or no integers at all, but it is at least a definite mathematical object which can be investigated. Second, if $Q = \{n : S(n)\}$, then the statements $Q = \mathbb{P}$ and $\forall n \in \mathbb{P} \ S(n)$ are equivalent. Thus, to prove that $S(n)$ is true for all positive integers n, form the set Q of all n for which it is true, establish that the hypotheses of Axiom 5 are valid, and conclude that $Q = \mathbb{P}$.

There are other sets satisfying the Peano axioms, for example the set $\{1, 3, 5, 7, ...\}$ of odd positive integers with $s(1) = 3, s(3) = 5, s(5) = 7, ...$ It can be proved that any two sets satisfying the Peano axioms are in bijection, and we view them as the same object (we call them isomorphic models).

You may wonder what happens if we drop some of the axioms. Do we still deal with the set of positive integers? For example, if we drop Axiom 3 and allow 1 to be in the range of s, we can take the finite set $\{1, 2, ..., n\}$ such that $s(1) = 2, s(2) = 3, ..., s(n) = 1$. For example, on a usual clock, 12 is followed by 1. This set with this successor function satisfies the other four axioms. It can be proved that Axiom 3 forces $\mathbb{P}$ to be infinite.

How do we know that $\mathbb{P}$ is not too big? If we drop the last axiom, we may consider $A = \mathbb{P} \cup \{\omega\}$, where $\omega \notin \mathbb{P}$ is a new element and we can extend s to A such that $s(\omega) = \omega$. The new set A satisfies the first four axioms, but not the last one (take $Q = \mathbb{P}$ which is a proper subset of A).

Definition 6.1. *We define the operation $+$ on $\mathbb{P}$ called* addition *such that*

$$m + 1 = s(m), \quad m + s(k) = s(m + k).$$

In other words, for a fixed m, we need to know how to define $m + 1$, and, once we know $m + k$, we define $m + s(k)$ as $s(m + k)$. Since any positive integer other than 1 is of the form $s(k)$ for some k, we are done. We say that we defined the addition inductively. This way, for each $m, n \in \mathbb{P}$ there is a unique $m + n \in \mathbb{P}$.

Example 6.2. *Let us show that $2 + 2 = 4$.*

Proof. We give a direct proof. Indeed, $2 + 2 = 2 + s(1)$ (definition of 2) $= s(2 + 1) = s(s(2))$ (definition of addition) $= s(3)$ (definition of 3) $= 4$ (definition of 4). $\qquad\square$

Theorem 6.3. *The addition of positive integers has the properties:*
    1. $\forall n \, (n + 1 \neq n)$.
    2. $\forall n \, (n = 1) \vee (\exists m(n = m + 1))$.
    3. $\forall m \forall n \forall p \, [(m + n) + p = m + (n + p)]$ *(addition is associative)*.

Proof. 1. This says that for all n, $s(n) \neq n$. This is true for $n = 1$, since $1 + 1 = 2 = s(1)$ cannot be equal to 1 by Axiom 3. Consider a k such that $s(k) \neq k$. Since the successor function is one-to-one (Axiom 4), we get $s(s(k)) \neq s(k)$, or $s(k + 1) \neq k + 1$. We proved by induction that for all $n \in \mathbb{P}$ we have $s(n) \neq n$.

    2. Given $n \in \mathbb{P} \setminus \{1\}$, we know that there is $m \in \mathbb{P}$ with $s(m) = n$, in other words, such that $n = m + 1$.

    3. We fix $m, n \in \mathbb{P}$ and use induction on p. For $p = 1$,

$$(m + n) + 1 = s(m + n) = m + s(n) = m + (n + 1).$$

Assume that $(m + n) + q = m + (n + q)$, and let's prove associativity for $p = s(q)$. We have

$$(m + n) + p = (m + n) + s(q) = s((m + n) + q) = s(m + (n + q)) =$$

$$= m + s(n + q) = m + ((n + q) + 1) = m + (n + (q + 1)) = m + (n + p).$$

$\qquad\square$

Lemma 6.4. $\forall n \, (n + 1 = 1 + n)$.

Proof. For $n = 1$ this is clear. Assume $k + 1 = 1 + k$ for a fixed k, and let $n = s(k)$. We have

$$1 + n = 1 + s(k) = s(1 + k) = s(k + 1) =$$

$$= k + s(1) = k + (1 + 1) = (k + 1) + 1 = s(k) + 1 = n + 1.$$

$\qquad\square$

Theorem 6.5. *The addition is commutative, more precisely*

$$\forall m \forall n \, (m + n = n + m).$$

Proof. We know this to be true for $n = 1$ by the previous lemma. Assume $m + k = k + m$, and let's prove it for $n = s(k)$. We have

$$m + s(k) = s(m + k) = s(k + m) = k + s(m) =$$

$$= k + (m + 1) = k + (1 + m) = (k + 1) + m = s(k) + m = n + m,$$

using the associativity of addition. $\qquad\square$

Theorem 6.6. *The addition of positive integers satisfies the cancellation law:*

$$\forall m \forall n \forall p \, (m + p = n + p \Rightarrow m = n).$$

Proof. We use induction on p. For $p = 1$, assuming $m + 1 = n + 1$, we get $s(m) = s(n)$, which implies $m = n$ since s is one-to-one. Assume that $m + k = n + k$ implies $m = n$ for some k, and let's prove that $m + s(k) = n + s(k) \Rightarrow m = n$. Since $m + s(k) = s(m + k)$ and $n + s(k) = s(n + k)$, from $s(m + k) = s(n + k)$ we obtain $m + k = n + k$, hence $m = n$. $\square$

6.2 The natural order relation and subtraction

Definition 6.7. *For $x, y \in \mathbb{P}$, we say that x is smaller than y and write $x < y$ if and only if there exists a positive integer u such that $x + u = y$. Other ways of expressing the relation $x < y$ are: x is less than y or y is greater than x.*

Recall that $\mathbb{P} = \{1, s(1), s(s(1)), s(s(s(1))), ...\}$. The relation $x < y$ can be understood like this: in the sequence $1, s(1), s(s(1)), ...$ the number x appears first, and y later. That is, y is obtained by applying s or s composed to s a number of times to x. As a consequence, $1 < 2 < 3 < ...$ and there is no positive integer m such that $x < m$ for all $x \in \mathbb{P}$.

Theorem 6.8. *The relation $<$ is transitive: $(x < y) \wedge (y < z) \Rightarrow x < z$.*

Proof. Since $x < y$ and $y < z$, there are u, v with $y = x + u$ and $z = y + v$. Then

$$z = y + v = (x + u) + v = x + (u + v),$$

hence $x < z$. $\square$

Lemma 6.9. *For $x, y \in \mathbb{P}$, we have*
  1. *$x = 1$ or $1 < x$.*
  2. *$x = y$ or $x < y$ or $y < x$ (law of trichotomy).*
  3. *$\neg(x < x)$.*

Proof. 1. If $x \neq 1$, we have seen that $x = s(k)$ for some k, hence $x = k + 1 = 1 + k$ and $1 < x$.

2. If $x \neq y$, suppose that in the sequence $1, s(1), s(s(1)), ...$ the number x appears first. This means that there is u such that $y = x + u$ and $x < y$. If y appears first, then $y < x$.

3. If $x < x$, we get that $x = x + u$ for some $u \in \mathbb{P}$, a contradiction. $\square$

Definition 6.10. *We proved above that the relation $<$ is transitive, hence it is a strict order relation. If we define $x \leq y$ if $x < y$ or $x = y$, then it is easy to prove that $\leq$ is reflexive, antisymmetric, and transitive, and hence an order relation. This is called the natural order relation on $\mathbb{P}$.*

Theorem 6.11. *For $x, y, z \in \mathbb{P}$ we have $x < y \Leftrightarrow x + z < y + z$.*

Proof. If $x < y$, there is u with $x + u = y$. Adding z we get $(x + u) + z = y + z$ or $(x + z) + u = y + z$, hence $x + z < y + z$. The converse uses cancellation. $\square$

Theorem 6.12. *For $x, y, z \in \mathbb{P}$ we have $x + y < z \Rightarrow (x < z) \wedge (y < z)$.*

Proof. Since $x + y < z$, we get $z = x + y + u$ for some u. In particular, $z = x + (y + u)$, hence $x < z$ and $z = y + (x + u)$, hence $y < z$. $\square$

Theorem 6.13. *For $x, y, z, u \in \mathbb{P}$ we have $(x < z) \wedge (y < u) \Rightarrow x + y < z + u$.*

Proof. Since $x < z$ and $y < u$ we get $z = x + v$ and $u = y + w$ for some v, w. Then $z + u = x + v + y + w = (x + y) + (v + w)$, therefore $x + y < z + u$. $\square$

Theorem 6.14. *For $x, y, z, u \in \mathbb{P}$ we have $x + y < z + u \Rightarrow (x < z) \vee (y < u)$.*

Proof. From $x + y < z + u$ we get $z + u = x + y + v$ for some v. If $x < z$, we are done. Assume $z \leq x$. Then $x = z$ or $x = z + w$ for some w. In the first case, by cancellation $u = y + v$, so $y < u$. In the second case, $z + u = z + w + y + v$ and $u = y + (v + w)$, so $y < u$ as well. $\square$

It is sometimes convenient to use the inverse of $<$, the greater-than relation, denoted $>$. This is defined symbolically by $x > y \Leftrightarrow y < x$. Clearly, any statement using $<$ has a natural counterpart using $>$ and all the theorems about *less than* can be converted into theorems about *greater than*. We can also define $x \geq y$ if $x = y$ or $x > y$. The relation $\geq$ is also reflexive, antisymmetric and transitive, so it is an order relation on $\mathbb{P}$.

Theorem 6.15. *(Archimedean property): For all $m, n \in \mathbb{P} \; \exists k \in \mathbb{P}$ with $m < kn$.*

Proof. If $m < n$, we can take $k = 1$. For $m = n$, $k = 2$ works. For $n < m$ we can take $k = m + 1$ since $m < mn + n$. $\square$

Theorem 6.16. *(Well-ordering principle) If A is a non-empty subset of $\mathbb{P}$, then A has a smallest element. More specifically, there is an element $x \in A$ such that $x \leq y$ for all $y \in A$. The smallest element is unique.*

Proof. Suppose there is a subset $A \neq \emptyset$ with no smallest element. We shall prove by induction on n that $x \in A \Rightarrow x \geq n$. For $n = 1$ this is obvious. Assume that the property holds for some $k \geq 1$. Then we cannot have $k \in A$, since this would be the smallest element. Hence $k \notin A$ and every $x \in A$ has the property that $x \geq k + 1$. By induction we got $x \in A \Rightarrow x \geq n$ for all $n \geq 1$. Since A is not empty, let $m \in A$ be some element. Using the inequality for $n = m + 1$ we get $m \geq m + 1$, a contradiction. It remains that all nonempty subsets have a smallest element.

Uniqueness follows from the fact that if $a, b \in A$ are both smallest elements, then $a \leq b$ and $b \leq a$ implies $a = b$. $\square$

Remark 6.17. *The well-ordering principle says that $\mathbb{P}$ with the natural order is a well-ordered set. This is not to be confused with Zermelo's well-ordering theorem, which is equivalent with the axiom of choice and Zorn's lemma discussed in the previous chapter. Warning: some authors call Zermelo's theorem the well-ordering principle.*

Theorem 6.18. *The well-ordering principle implies the principle of mathematical induction.*

Proof. Indeed, assume that for each positive integer n, a statement $S(n)$ is given. We assume that $S(1)$ is true, and that $S(k) \Rightarrow S(k+1)$. Let's prove that $S(n)$ is true for all n. If not, let $Q \neq \emptyset$ be the subset of $\mathbb{P}$ consisting of those m for which $S(m)$ is false. By the well-ordering principle, Q contains a smallest element d. Since $S(1)$ is true, we must have that $d > 1$. But then there is e with $d = s(e) = e + 1$ and e cannot be in Q, so $S(e)$ must be true. From this hypothesis, we get $S(e+1) = S(d)$ true, a contradiction. Therefore Q must be empty and $S(n)$ is true for all n. $\qquad\square$

Lemma 6.19. *For any $x, y \in \mathbb{P}$ with $x < y$ there exists a unique $z \in \mathbb{P}$ such that $x + z = y$.*

Proof. Recall that since $x < y$, there is z with $x + z = y$. If z' also satisfies $x + z' = y$, we get $x + z = x + z'$, so $z = z'$ by cancellation. $\qquad\square$

Definition 6.20. *Suppose that $x, y \in \mathbb{P}$ with $x < y$. We define the subtraction operation and write $y - x$ to denote the unique $z \in \mathbb{P}$ such that $x + z = y$. We say that we have subtracted x from y. In other words, $y - x$ stands for the positive integer such that $x + (y - x) = y$.*

Example 6.21. *Let's prove that*
1. $4 - 2 = 2$.
2. $(x + y) - x = y$.
3. $[x + (y + z)] - z = x + y$.
4. $x + y < z \Rightarrow y < z - x$.
5. $(x < y) \wedge (y < z) \Rightarrow y - x < z$.

Proof. Indeed, for 1 we have seen that $2 + 2 = 4$. Equation 2 follows since $x + y > x$ and $x + y = x + y$. Part 3 uses associativity of addition. For 4, notice that $z = x + y + u$ and therefore $z - x = y + u$. For 5, let $y = x + u, z = y + v$ for some $u, v \in \mathbb{P}$. Then $u = y - x$ and $z = x + u + v$, in particular $u < z$. $\quad\square$

Theorem 6.22. *For $x, y, z \in \mathbb{P}$ we have*
1. $z < y \Rightarrow (x + y) - z = x + (y - z)$.
2. $x + y < z \Rightarrow z - (x + y) = (z - x) - y$.
3. $(x < y) \wedge (y < z) \Rightarrow z - (y - x) = (z - y) + x$.
4. $(x < y) \wedge (z < x) \Rightarrow x - z < y - z$.
5. $(x < y) \wedge (y < z) \Rightarrow z - y < z - x$.

Proof. 1. Since $z < y$, we have $z+(y-z) = y$. Adding x we get $(z+(y-z))+x = y + x$ or $((y - z) + x) + z = x + y$, hence $(y - z) + x = (x + y) - z$.

2. Note that $x + y < z$ implies $x < z$ and $y < z - x$. Using part 1 and the fact that $(x+y) - x = y$, we get $(x+y) + [(z-x) - y] = (x+y+z - x) - y = (y + z) - y = z$.

3. We have $[(z-y) + x] + (y-x) = (z-y) + [x + (y-x)] = (z-y) + y = z$, so $(z - y) + x = z - (y - x)$.

4. Since $z < x < y$, we get $z < y$. We have $y - z = (y - x) + (x - z)$, hence $x - z < y - z$.

5. Again $x < z$, and $z - x = (z - y) + (y - x)$, so $z - y < z - x$. □

6.3   Multiplication and divisibility

Definition 6.23. *We inductively define the operation of* multiplication *(denoted by a dot $\cdot$, which is sometimes omitted) on $\mathbb{P}$ as follows:*

$$x \cdot 1 = x \quad and \quad x \cdot s(y) = x \cdot (y + 1) = x \cdot y + x.$$

Theorem 6.24. *Multiplication on $\mathbb{P}$ has these properties:*
1. $x \cdot (y + z) = x \cdot y + x \cdot z$ and $(x + y) \cdot z = x \cdot z + y \cdot z$ (distributivity).
2. $x \cdot y = y \cdot x$ (commutativity).
3. $x \cdot (y \cdot z) = (x \cdot y) \cdot z$ (associativity).
4. $x < y \Rightarrow x \cdot z < y \cdot z$.
5. $x \cdot z = y \cdot z \Rightarrow x = y$ (cancellation).
6. $x \cdot z < y \cdot z \Rightarrow x < y$.
7. $x < y \Rightarrow z \cdot (x - y) = z \cdot y - z \cdot x$.
8. $x < y \wedge u < v \Rightarrow x \cdot u < y \cdot v$.
9. $y \neq 1 \Rightarrow x < x \cdot y$.
10. $x \leq x \cdot y$.

Proof. 1. By induction on z. For $z = 1$,

$$x \cdot (y + 1) = x \cdot y + x = x \cdot y + x \cdot 1.$$

Assume $x \cdot (y + k) = x \cdot y + x \cdot k$. Then

$$x \cdot (y + (k + 1)) = x \cdot [(y + k) + 1] = x \cdot (y + k) + x =$$

$$= x \cdot y + x \cdot k + x \cdot 1 = x \cdot y + x \cdot (k + 1).$$

The other equality is proved similarly.

2. First we prove by induction on x that $1 \cdot x = x$. This is certainly true for $x = 1$. Assume $1 \cdot k = k$. Then

$$1 \cdot (k + 1) = 1 \cdot k + 1 \cdot 1 = k + 1.$$

Now we use induction on y to show that $x \cdot y = y \cdot x$. This is true for $y = 1$ since $x \cdot 1 = x = 1 \cdot x$. Assume $x \cdot k = k \cdot x$. Then

$$x \cdot (k + 1) = x \cdot k + x = k \cdot x + x = (k + 1) \cdot x.$$

3. By induction on z we have

$$x \cdot (y \cdot 1) = x \cdot y = (x \cdot y) \cdot 1.$$

Assume $x \cdot (y \cdot k) = (x \cdot y) \cdot k$. Then

$$x \cdot (y \cdot (k + 1)) = x \cdot (y \cdot k + y) = x \cdot (y \cdot k) + x \cdot y =$$
$$= (x \cdot y) \cdot k + x \cdot y = (x \cdot y) \cdot (k + 1).$$

4. For $z = 1$ it is true that $x < y \Rightarrow x < y$. Assume that $x < y \Rightarrow x \cdot k < y \cdot k$ for some k, and let's prove it for $k + 1$. By adding the inequalities $x < y$ and $x \cdot k < y \cdot k$ we get $x \cdot (k + 1) < y \cdot (k + 1)$.

5. We prove the contrapositive: $x \neq y \Rightarrow x \cdot z \neq y \cdot z$. By trichotomy, if $x \neq y$, there are two possibilities: $x < y$ or $x > y$. In the first case we get $x \cdot z < y \cdot z$ and in the second case $x \cdot z > y \cdot z$, hence $x \cdot z \neq y \cdot z$.

We leave the other properties as an exercise. $\square$

Definition 6.25. *Given $x, y \in \mathbb{P}$, we say that x divides y, x is a* factor *or* divisor *of y or y is a* multiple *of x if and only if $\exists u$ such that $x \cdot u = y$. We write this symbolically as $x \mid y$, and we obtain the* divisibility *relation on $\mathbb{P}$.*

Theorem 6.26. *We have*
  1. $(x \mid y) \wedge (y \mid z) \Rightarrow x \mid z$ (transitivity).
  2. $(x \mid y) \wedge (y \mid x) \Rightarrow x = y$ (antisymmetry).
  3. $x \mid y \Rightarrow x \leq y$.
  4. $(1 \mid x) \wedge (x \mid x)$.
  5. $x \mid y \Leftrightarrow x \cdot z \mid y \cdot z$.
  6. $(x \mid y) \wedge (x \mid z) \Rightarrow x \mid (y + z)$.
  7. $(x \mid y) \wedge (x \mid z) \wedge (y < z) \Rightarrow x \mid (z - y)$.

Proof. 1. We have $y = x \cdot u$ and $z = y \cdot v$ for some u, v. It follows that $z = x \cdot (uv)$, hence $x \mid z$.

2. We have $y = x \cdot u$ and $x = y \cdot v$, hence $y = y \cdot (uv)$. It follows that $uv = 1$, so $u = v = 1$ and $x = y$.

3. Since $y = x \cdot u$ for some u and $x \leq x \cdot u$, we get $x \leq y$.

4. We have $x = 1 \cdot x = x \cdot 1$.

5. $\Rightarrow$: Suppose $x \mid y$. Then there exists a positive integer u such that $x \cdot u = y$. Multiplying by z we get $(x \cdot u) \cdot z = y \cdot z$. Using associativity and commutativity on the left side, we can write the equation in the form $(x \cdot z) \cdot u = y \cdot z$. Thus by Definition 6.25 we have $x \cdot z \mid y \cdot z$.

$\Leftarrow$: If $x \cdot z \mid y \cdot z$, then for some u, $(x \cdot z) \cdot u = y \cdot z$. As above, this can be rewritten as $(x \cdot u) \cdot z = y \cdot z$, and by multiplicative cancellation, $x \cdot u = y$. Thus $x \mid y$.

We leave parts 6 and 7 as exercises. $\square$

Corollary 6.27. *Divisibility is an order relation on* $\mathbb{P}$.

Proof. Indeed, it is reflexive, antisymmetric, and transitive. □

Unlike the relation $\leq$, the order relation $|$ is not a total order. For example, $2 \nmid 3$ and $3 \nmid 2$, so these two elements are incomparable.

Recall that given $x, y \in \mathbb{P}$ with $y \mid x$, there exists a unique $u \in \mathbb{P}$ such that $x = y \cdot u$. Indeed, if $x = y \cdot u = y \cdot v$, we get $u = v$ by part 4 in Theorem 6.24.

Definition 6.28. *If* $y \mid x$, *then the unique* u *such that* $y \cdot u = x$ *is denoted by* $x \div y$ *and is called the* quotient *of* x *by* y. *This defines the* division *operation for selected positive integers. Another notation for* $x \div y$ *is* x/y *or* $\dfrac{x}{y}$. *Note that:* $y \mid x \Rightarrow y \cdot (x \div y) = x$ *and* $y \cdot u = x \Rightarrow u = x \div y$.

Theorem 6.29. *The division operation has these properties:*

1. $(y \mid x) \wedge (z \mid y) \Rightarrow (y \div z) \mid (x \div z)$.
2. $z \mid y \Rightarrow (x \cdot y) \div z = x \cdot (y \div z)$.
3. $(y \mid x) \wedge (x \mid z) \Rightarrow (z \div x) \mid (z \div y)$.
4. $x \cdot y \mid z \Rightarrow z \div (x \cdot y) = (z \div x) \div y$.
5. $(x \mid y) \wedge (x \mid z) \Rightarrow (y + z) \div x = (y \div x) + (z \div x)$.
6. $(x \mid y) \wedge (x \mid z) \wedge (z < y) \Rightarrow (y - z) \div x = (y \div x) - (z \div x)$.

Proof. 1. We have $x = y \cdot u$ and $y = z \cdot v$ for some u, v, so $x = z \cdot uv$. Moreover, $y \div z = v$ and $x \div z = uv$, hence $(y \div z) \mid (x \div z)$.

2. Since $z \mid y$, we have $y = zu$ for some u, hence $y \div z = u$. We get $xy = xzu$ and $(x \cdot y) \div z = x \cdot u$.

We leave the other parts as an exercise. □

6.4 Natural numbers

The set of natural numbers $\mathbb{N} = \mathbb{P} \cup \{0\}$ satisfies the following axioms:

Axiom 1. There exists an element $0 \in \mathbb{N}$.

Axiom 2. For each $n \in \mathbb{N}$, there is an element $s(n) \in \mathbb{N}$, called the successor of n.

Axiom 3. For each $n \in \mathbb{N}$ we have $s(n) \neq 0$.

Axiom 4. If $m, n \in \mathbb{N}$ and $s(m) = s(n)$, then $m = n$.

Axiom 5. If A is a subset of $\mathbb{N}$ such that $0 \in A$ and $s(n) \in A$ whenever $n \in A$, then $A = \mathbb{N}$.

The number 0 was discovered much later, and it could be thought of as the number of elements in the empty set. Axiom 5 implies the principle of mathematical induction for $\mathbb{N}$: If $P(n)$ is a statement for each $n \in \mathbb{N}$ and we prove that

1. $P(0)$ true, and

2. $P(k)$ true implies $P(k+1)$ true for all $k \geq 0$, then $P(n)$ is true for all $n \in \mathbb{N}$.

We can extend the addition and multiplication operations to the set of natural numbers, with the new rules $x + 0 = x$, $x \cdot 0 = 0$. When we define divisibility, we exclude 0 as a divisor.

We list the major properties of the natural numbers, especially those that we will use later when we study the integers. All the proofs are similar to the ones for positive integers.

Theorem 6.30. *For the set* $\mathbb{N} = \{0, 1, 2, 3, ...\}$ *of natural numbers we have:*

1) Addition and multiplication are commutative and associative.

2) Multiplication is distributive with respect to addition.

3) We have cancelation properties: $x + y = x + z \Rightarrow y = z$ *and* $(x \cdot y = x \cdot z \wedge x \neq 0) \Rightarrow y = z$.

4) There is a natural order defined by $x < y \Leftrightarrow \exists u \in \mathbb{N} \setminus \{0\}$ *such that* $x + u = y$ *and we define* $x \leq y \Leftrightarrow (x < y) \vee (x = y)$.

5) The trichotomy property holds: $\forall x, y \in \mathbb{N}$, *exactly one of the following is true:* $x < y$, $x = y$, *or* $y < x$.

6) We have the Archimedean property: $\forall m, n \in \mathbb{N}$ *with* $n \neq 0$ *there exists* $k \in \mathbb{P}$ *with* $m < kn$.

7) We have $x < y \Leftrightarrow x + z < y + z$ *for all* $z \in \mathbb{N}$ *and if* $z \neq 0$, *then* $x < y \Leftrightarrow xz < yz$.

8) The well-ordering principle is: If $A \subseteq \mathbb{N}$, $A \neq \emptyset$, *then* A *contains a unique smallest element.*

Definition 6.31. *For* $x \in \mathbb{P}$ *and* $y \in \mathbb{N}$ *we define the* power x^y *inductively by* $x^0 = 1$, $x^{y+1} = x^y \cdot x$. *The number* x *is called the* base *and* y *is called the* exponent.

Definition 6.32. *For* $n \in \mathbb{N}$, *the factorial function* $n!$ *is defined by* $0! = 1$ *and* $(n+1)! = (n+1) \cdot n!$.

Note that $1! = 1, 2! = 2, 3! = 6, 4! = 24$, etc. and for $m \geq 1$ we have $m \leq n \Rightarrow m \mid n!$.

6.5   Other forms of induction

In this section we give the justification for the method of strong or complete induction with applications to arithmetic. We also illustrate the induction with bigger steps.

Theorem 6.33. *(Strong induction or complete induction) If* A *is a subset of* $\mathbb{N}$ *such that*

1. $0 \in A$ *and*
2. $\forall n \geq 1, \{0, 1, 2, ..., n\} \subseteq A \Rightarrow n + 1 \in A$,

then $A = \mathbb{N}$.

Proof. Consider the statement $0, 1, 2, ..., n \in A$, denoted $P(n)$. Since $0 \in A$, $P(0)$ is true. Assume $P(k)$ is true, so $0, 1, 2, ..., k \in A$. From part 2 we get that $k + 1 \in A$, hence $P(k + 1)$ is true. By induction, it follows that $P(n)$ is true for all n, in particular $A = \mathbb{N}$. $\qquad\square$

Note that, in fact, strong induction implies the usual induction, hence they are equivalent. Using strong induction and generalized induction, we get

Corollary 6.34. *If A is a subset of $\mathbb{N}$ such that $k_0 \in A$ and $(m \leq n \Rightarrow m \in A) \Rightarrow n + 1 \in A$, then $A = \{k_0, k_0 + 1, ...\}$.*

Definition 6.35. *A positive integer x is a prime if and only if $x \neq 1$ and $y \mid x$ implies $(y = 1) \vee (y = x)$.*

Theorem 6.36. *1. A positive integer x is prime if and only if $x \neq 1$ and*

$$\forall r \forall s (x = r \cdot s \Rightarrow (r = 1) \vee (r = x)).$$

2. If $y \neq 1$, then there is a prime p such that $p \mid y$.

Proof. The first part follows directly from the definition. For part 2 we use complete induction. If $y = 2$, then 2 is a prime and $2 \mid y$. Assume the statement is true for $2, 3, ..., y$ and let's prove it for $y + 1$. If $y + 1$ is prime, we are done. If not, then it has a divisor $2 \leq d \leq y$. Since d has a prime divisor, it follows that $y + 1$ also has a prime divisor. $\qquad\square$

It is worth noting that, in the set of positive integers, a prime is a number which has exactly two divisors: 1 and itself. However, in the larger set $\mathbb{Z}$ consisting of all integers, we will see that a prime is a number p with exactly four divisors: $\pm 1, \pm p$.

Theorem 6.37. *Each positive integer $n \geq 2$ is either a prime or is a product of primes.*

Proof. The basis step is true, since 2 is a prime. Suppose that the integers $2, 3, ..., k$ are either primes or product of primes, and let's prove that $k + 1$ also has this property. If it happens that $k + 1$ is a prime, we are done. If not, then $k + 1 = a \cdot b$, where $2 \leq a, b \leq k$. By hypothesis, both a and b are either primes or products of primes. It follows that $k + 1 = a \cdot b$ is a product of primes as well, and we are done. $\qquad\square$

Theorem 6.38. *A positive integer n is prime if and only if n is not divisible by any prime p with $2 \leq p \leq \sqrt{n}$.*

Proof. If n is prime, then the only divisors are 1 and n, so the implication $\Rightarrow$ is clear. For the converse, assume that n is not prime. Then n has a divisor $d \neq 1, n$ and $n = dm$ where $d \leq \sqrt{n}$ or $m \leq \sqrt{n}$. They cannot both be bigger than $\sqrt{n}$, since in that case $dm > \sqrt{n}\sqrt{n} = n$. By taking a prime factor of d or m, we are done. $\qquad\square$

Theorem 6.39. *(Induction with bigger steps) Consider a statement $P(n)$ for each $n \geq k_0$. Suppose*
  1. $P(k_0), P(k_0 + 1), ..., P(k_0 + k_1 - 1)$ *are true for a fixed integer $k_1 \geq 1$, called a step, and*
  2. $P(k)$ *true implies $P(k + k_1)$ true for all $k \geq k_0$.*
  Then $P(n)$ is true for all $n \geq k_0$.

Proof. Indeed, by the division algorithm (see next chapter for a proof), any integer $n \geq k_1$ is of the form $n = qk_1 + r$ where $q \geq 1$ and $0 \leq r \leq k_1 - 1$. $\quad\square$

Example 6.40. *For $n \geq 6$, any square can be partitioned into n squares using segments parallel with its sides.*

Proof. Obviously a square can be partitioned into 4 equal squares using segments through the midpoints of the sides.

We prove now that we can partition a square into $6, 7$, or 8 squares, and then we will prove the induction step from k to $k+3$. Here $k_0 = 6$ and $k_1 = 3$. Indeed, the following picture illustrates the cases $n = 6, 7$, and 8:

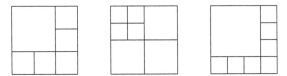

Suppose we know how to divide a square into k squares for $k \geq 6$. Take one of the small squares and divide it into four squares. This way, we get a partition of the initial square into $k + 3$ squares. $\qquad\square$

6.6   Exercises

Exercise 6.41. *Prove the following properties of the natural order relation $\leq$ for $x, y, z \in \mathbb{P}$.*

   a) $x \leq y \Leftrightarrow x + z \leq y + z$.
   b) $(x \leq y) \vee (y \leq x)$.
   c) $y \leq x \Leftrightarrow y < x + 1$.

Exercise 6.42. *Prove that the set $E = \{0, 2, 4, 6, ...\}$ of even natural numbers is another set satisfying the Peano axioms for $\mathbb{N}$, defining an appropriate successor function.*

Exercise 6.43. *For $n \in \mathbb{N}$, define the set of descendants $D(n)$ as the smallest subset of $\mathbb{N}$ such that $n \in D(n)$ and it is closed under the successor function: $m \in D(n) \Rightarrow s(m) \in D(n)$. Prove the following properties:*

   1. $D(n) = \{n\} \cup D(s(n))$.
   2. $D(s(n)) = s(D(n))$.
   3. $n \notin D(s(n))$.
   4. $D(m) = D(n) \Rightarrow m = n$.
   5. If $\emptyset \neq A \subseteq \mathbb{N}$ and A is closed under s, in the sense that $s(A) \subseteq A$, then there is a unique $k \in \mathbb{N}$ such that $A = D(k)$.
   6. For $m, n \in \mathbb{N}, m \leq n \Leftrightarrow n \in D(m)$.

Exercise 6.44. *Prove that for $x \in \mathbb{P}$ and $y, z \in \mathbb{N}$,*

   a. $x^{y+z} = x^y \cdot x^z$.
   b. $x^{yz} = (x^y)^z$.

Exercise 6.45. *Suppose $u_0 = 2, u_1 = 3$ and $u_{n+1} = 3u_n - 2u_{n-1}$ for all $n \geq 1$. Prove by strong induction that $u_n = 2^n + 1$.*

Exercise 6.46. *Prove that any positive integer n can be written as a product of an odd integer and a power of 2.*

Exercise 6.47. *Let B be a set with $n \geq 3$ elements. Assume we know that the number of subsets with two elements is $n(n-1)/2!$. Prove that the number of subsets with three elements is $n(n-1)(n-2)/3!$.*

Exercise 6.48. *Prove, by induction, that it is possible to pay, with 3-rouble and 5-rouble banknotes, any whole number $n \geq 8$ of roubles without requiring that the payer receive any change.*

Exercise 6.49. *"Prove that any cube can be partitioned into $n \geq 58$ cubes using planes parallel with its faces. (Hint: Use the fact that a cube can be partitioned into 8 cubes and into 27 cubes to verify the statement for $58, 59, ..., 64$ and then use induction with step 7).*

7

Elementary number theory

In this chapter we discuss divisibility properties of the whole numbers or integers $\ldots -3, -2, -1, 0, 1, 2, 3, \ldots$. We prove the fundamental theorem of arithmetic which says that any nonzero integer different from ± 1 is a prime or a product of primes in a unique way up to a permutation of factors. We also discuss the greatest common divisor and least common multiple of two integers. We prove the division algorithm and the Euclidean algorithm. We develop several criteria of divisibility. The axiomatic construction of the set of integers $\mathbb{Z}$ is postponed to Chapter 10.

7.1 Absolute value and divisibility of integers

Definition 7.1. *If x is any integer, then the absolute value of x is the natural number defined by*

$$|x| = \begin{cases} x & \text{if } x \geq 0 \\ -x & \text{if } x < 0. \end{cases}$$

Theorem 7.2. *We have the following properties of the absolute value:*
1. $|-x| = |x|$.
2. $x \leq |x|$ and $-x \leq |x|$.
3. $|x| = |y| \Leftrightarrow (x = y) \vee (x = -y)$.
4. $|xy| = |x| \cdot |y|$.
5. $a > 0 \Rightarrow (|x| \leq a \Leftrightarrow -a \leq x \leq a)$.
6. $|x| < |y| \Leftrightarrow -|y| < x < |y|$.
7. $|x + y| \leq |x| + |y|$ *(triangle inequality)*.
8. $|x - y| \leq |x| + |y|$.
9. $|x| - |y| \leq |x - y|$.

Proof. 1 and 2 follow from the definition, considering the cases $x \geq 0$ and $x < 0$.

3. There are four cases to consider: $x \geq 0$ and $y \geq 0$, $x \geq 0$ and $y < 0$, $x < 0$ and $y \geq 0$, $x < 0$ and $y < 0$. In each case we get $x = y, x = -y, -x = y$ and $-x = -y$, respectively.

4. We consider the four cases as in 3 and compute.

5. For $x \geq 0$ we get $x \leq a$. For $x < 0$ we get $-x \leq a$, hence $x \geq -a$.

6. We take $a = |y|$ and use 5.

7. If $x \geq 0$ and $y \geq 0$ this is clear. If $x \geq 0$ and $y < 0$, there are two subcases $x + y \geq 0$ and $x + y < 0$. In the first subcase, $x + y \leq x - y$ is true since $y < -y$. In the second subcase, $-x - y \leq x - y$ is true since $-x \leq x$. If $x < 0$ and $y \geq 0$, we treat similarly the subcases $x + y \geq 0$ and $x + y < 0$. If $x < 0$ and $y < 0$, then $-x - y \leq -x - y$ is clearly true.

8. We write $x - y = x + (-y)$, use 7 and 1.

9. We have $|x| = |x - y + y| \leq |x - y| + |y|$, hence $|x| - |y| \leq |x - y|$. $\qquad \square$

Definition 7.3. *Let x, y be integers. We say that y is divisible by x, x divides y, or x is a factor of y if and only if $x \neq 0$ and there exists an integer z such that $x \cdot z = y$. We will write $x \mid y$ when x divides y. The integer z is also denoted by y/x or $\dfrac{y}{x}$.*

Note that $x \mid 0$ for any nonzero integer x and $0/x = 0$.

Theorem 7.4. *If x and y are nonzero integers, then:*

1. $x \mid y \Rightarrow |x| \leq |y|$.
2. $x \mid y \Rightarrow |x| \, | \, |y|$.
3. $x \mid 1 \Rightarrow x = \pm 1$.
4. $(x \mid y) \wedge (y \mid x) \Rightarrow x = \pm y$.
5. $(x \mid y) \wedge (y \mid z) \Rightarrow x \mid z$ *(transitivity)*.
6. $(x \mid y) \wedge (x \mid z) \Rightarrow x \mid (y \pm z)$.

Proof. 1. If $x \mid y$, then $y = xa$ for $a \in \mathbb{Z} \setminus \{0\}$ and $|y| = |xa| = |a||x| \geq |x|$ since $|a| \geq 1$.

2. As above, if $x \mid y$, then $|y| = |a||x|$, hence $|x| \, | \, |y|$.

3. If $1 = ax$, then $a = x = 1$ or $a = x = -1$.

4. Assuming $x \mid y \wedge y \mid x$, we get $y = ax$ and $x = by$ for some integers a, b. It follows that $y = aby$, hence $ab = 1$ and therefore $a = b = 1$ or $a = b = -1$. In the first case $y = x$, and in the second $y = -x$.

5. Assuming $x \mid y \wedge y \mid z$, we get $y = ax$ and $z = by$, hence $z = abx$ and $x \mid z$.

6. Assuming $x \mid y \wedge x \mid z$, let $y = ax$ and $z = bx$. Then $y \pm z = ax \pm bx = (a \pm b)x$, hence $x \mid y \pm z$. $\qquad \square$

Remark 7.5. *The divisibility relation $\mid$ on $\mathbb{Z} \setminus \{0\}$ is reflexive and transitive, but not symmetric because $-2 \mid 4$ and $4 \nmid -2$ or antisymmetric because $-2 \mid 2$ and $-2 \neq 2$.*

Theorem 7.6. *(Division algorithm) For any integer y and any nonzero integer x there exist unique integers q and r such that $y = qx + r$ and $0 \leq r < |x|$.*

Proof. Assume first that $x > 0$. Consider the set S of natural numbers of the form $y - kx$ where $k \in \mathbb{Z}$. Notice that S is not empty: for example $y + |y|x \in S$.

Indeed, since $x \geq 1$ and $|y| \geq -y$, we get $y + |y|x \geq 0$ and we can take $k = -|y|$. By the well-ordering principle, S contains a smallest element of the form $r = y - qx$. We found integers q and r with $y = qx + r$ and we know that $r \geq 0$. To show that $r < x$, by way of contradiction, assume that $r \geq x$. Then $r - x \geq 0$ and $r - x = y - qx - x = y - (q+1)x \in S$. We found an element in S, namely $r - x$, such that $r - x < r$, which contradicts the fact that r was the smallest. Hence $r < x$.

To show uniqueness, suppose that q' and r' are some integers such that $y = q'x + r'$ and $0 \leq r' < x$. We will show that $q' = q$ and $r' = r$. From $qx + r = q'x + r'$ we get $(q - q')x = r' - r$. By adding the inequalities $-x < -r \leq 0$ and $0 \leq r' < x$, we obtain $-x < r' - r < x$, so using $r' - r = (q - q')x$ we get $-x < (q - q')x < x$. Canceling x, we obtain $-1 < q - q' < 1$, which means that $q - q' = 0$ or $q = q'$. Substituting $q = q'$ in the equation $r' - r = (q - q')x$, we get $r = r'$.

It remains to prove the theorem for $x < 0$. Let $x' = |x| = -x > 0$. Applying the division algorithm for y and x', we find unique q_1 and r such that $y = q_1 x' + r$ and $0 \leq r < x'$. Take $q = -q_1$. We conclude that $y = qx + r$ with $0 \leq r < |x|$, and q, r are unique. $\qquad\square$

7.2   Greatest common divisor and least common multiple

Definition 7.7. *An integer $z \neq 0$ is a* common divisor *of the integers x and y if and only if $z \mid x$ and $z \mid y$.*

An integer z is a common multiple *of x and y if and only if $x \mid z$ and $y \mid z$.*

Definition 7.8. *We say that z is a* greatest common divisor *of x and y if and only if z is a common divisor of x and y, and for all t such that $t \mid x$ and $t \mid y$, it follows that $t \mid z$. For $x, y \neq 0$, the positive greatest common divisor of x and y is denoted by $\gcd(x, y)$. Note that $\gcd(0, 0)$ is undefined, and for nonzero x, $\gcd(x, 0) = \gcd(0, x) = |x|$.*

We say that z is a least common multiple *of x and y if and only if z is a common multiple of x and y and for all t such that $x \mid t$ and $y \mid t$ it follows that $z \mid t$. The positive least common multiple of x and y is denoted by $\mathrm{lcm}(x, y)$. If one of x and y is zero, then $\mathrm{lcm}(x, y)$ is undefined.*

Example 7.9. $\gcd(4, 6) = 2$, $\mathrm{lcm}(4, 6) = 12$, $\gcd(30, -45) = 15$, $\mathrm{lcm}(30, -45) = 90$, $\gcd(-4, -33) = 1$, $\mathrm{lcm}(-4, -33) = 132$.

Theorem 7.10. *If $\emptyset \neq A \subseteq \mathbb{Z}$ and A is closed under subtraction, then there exists an element $d \in A$ such that $A = \{x \cdot d : x \in \mathbb{Z}\}$.*

Proof. If $A = \{0\}$, we can take $d = 0$. Assume that A has nonzero elements. Consider $S = \{|x| : x \in A, x \neq 0\}$. Then $S \subseteq \mathbb{P}$ is not empty, therefore has a least element $a > 0$. Note that $a \in A$. For any $x \in A$ there are q, r such that $x = qa + r$ and $0 \leq r < a$. Since $r = x - qa$ and A is closed under subtraction, it follows that $r \in A$. It must be that $r = 0$ and $x = qa$, since a was the smallest positive element and $r < a$. We can take $d = a$ and we conclude $A = \{q \cdot d : q \in \mathbb{Z}\}$. $\qquad\square$

Theorem 7.11. *If x and y are nonzero integers, then $\gcd(x, y)$ and $\mathrm{lcm}(x, y)$ exist. Moreover, there are $a, b \in \mathbb{Z}$ such that $\gcd(x, y) = ax + by$.*

Proof. Consider the set $S = \{xm + yn : m, n \in \mathbb{Z}\}$. Note that $1 \leq x \cdot x + y \cdot y \in S$, hence $S \cap \mathbb{P} \neq \emptyset$. By the well-ordering principle, $S \cap \mathbb{P}$ contains a smallest element $t \geq 1$. We claim that $t = \gcd(x, y)$. Let's show first that $t \mid x$.

By the division algorithm, there are $q, r \in \mathbb{Z}$ such that $x = tq + r$ with $0 \leq r < t$. We know that $t \in S$ and that there are $a, b \in \mathbb{Z}$ such that $t = ax + by$. By an easy computation, it follows that

$$r = x - tq = x - (ax + by)q = (1 - aq)x + (-bq)y,$$

hence $r \in S$. Since $r < t$, the only possibility is $r = 0$, hence $x = tq$ and $t \mid x$.

A similar argument shows $t \mid y$, hence t is a common divisor. Let z be another common divisor, and let $x = zu, y = zv$. Then $t = ax + by = auz + bvz = (au + bv)z$, hence $z \mid t$. It follows that $t = \gcd(x, y)$.

Since there is at least one positive common multiple for each pair of nonzero integers x, y (for example $|xy|$), by the well-ordering principle the set of positive common multiples has a smallest element. This proves that $\mathrm{lcm}(x, y)$ exists. $\qquad\square$

Lemma 7.12. *Let a, b, c be integers such that $a \mid bc$ and $\gcd(a, b) = 1$. Then $a \mid c$.*

Proof. Let $bc = ad$. Since $\gcd(a, b) = 1$, there are integers u, v such that $1 = au + bv$. Therefore $c = c \cdot 1 = cau + cbv = cau + adv = a(cu + dv)$ and $a \mid c$. $\qquad\square$

Lemma 7.13. *Let x, y be nonzero integers. If $\gcd(x, y) = 1$, then $\mathrm{lcm}(x, y) = |xy|$.*

Proof. Let $m = |xy|$. It is clear that $x \mid m$ and $y \mid m$, so m is a common multiple. Suppose $x \mid k$ and $y \mid k$. Then $k = xk'$ for some integer k'. Since $y \mid xk'$ and $\gcd(x, y) = 1$, by Lemma 7.12 we get $y \mid k'$. It follows that $|xy| = m \mid k$, hence $\mathrm{lcm}(x, y) = |xy|$. $\qquad\square$

Corollary 7.14. *For nonzero integers x, y we have $\mathrm{lcm}(x, y) \cdot \gcd(x, y) = |xy|$.*

Proof. Let $d = \gcd(x, y)$. Then $x = dx', y = dy'$ for some integers x', y' and $\gcd(x', y) = 1$. We will prove that $\mathrm{lcm}(x, y) = \frac{|xy|}{d}$. Let $m = \frac{|xy|}{d} = \pm x'y = \pm xy'$. It is clear that $x \mid m$ and $y \mid m$, so m is a common multiple. Assume $x \mid k$ and $y \mid k$. Then $x' \mid k$ and $y \mid k$, so as in Lemma 7.13 we conclude that $x'y \mid k$, so $m \mid k$. It follows that $m = \mathrm{lcm}(x, y)$. $\qquad\square$

Exercise 7.15. *Prove that* $gcd(6n + 8, 4n + 5) = 1$.

Solution. Assume $d \mid 6n+8$ and $d \mid 4n+5$. Then $d \mid 2(6n+8) - 3(4n+5) = 1$.

Theorem 7.16. *We have the properties:*
1. $(x \mid z) \wedge (y \mid z) \Rightarrow lcm(x, y) \mid z$.
2. $lcm(lcm(x, y), z) = lcm(x, lcm(y, z))$ *(associativity)*.
3. $z \mid x \wedge z \mid y \Rightarrow z \mid gcd(x, y)$.
4. $gcd(gcd(x, y), z) = gcd(x, gcd(y, z))$ *(associativity)*.
5. $x \mid y \Rightarrow gcd(x, y) = |x|$.
6. $y = q \cdot x + r \Rightarrow gcd(x, y) = gcd(r, x)$.

Proof. Parts 1 and 3 follow from the definition.

2. Both are equal to $lcm(x, y, z)$, the least common multiple of x, y, z, defined as the smallest positive integer divisible by x, y, z.

4. Both sides are equal to $gcd(x, y, z)$, the largest positive integer that divides x, y, z.

5. Since $|x| \mid x$ and $|x| \mid y$, we get $|x| \le gcd(x, y)$. Since also $gcd(x, y) \le |x|$, we get equality.

6. If d divides x and y, then $d \mid y - q \cdot x = r$, hence $d \mid gcd(r, x)$. Conversely, if $d \mid r$ and $d \mid x$, then $d \mid q \cdot x + r = y$, hence $d \mid gcd(x, y)$. By double inequality, we get $gcd(x, y) = gcd(r, x)$. $\qquad\square$

Theorem 7.17. *(Euclidean algorithm) Let a, b be positive integers with $a \ge b$. If $b \mid a$, then $gcd(a, b) = b$. If $b \nmid a$, apply the division algorithm repeatedly as follows:*

$$a = bq_0 + r_0, \quad 0 < r_0 < b$$

$$b = r_0 q_1 + r_1, \quad 0 \le r_1 < r_0$$

$$r_0 = r_1 q_2 + r_2, \quad 0 \le r_2 < r_1$$

$$\vdots$$

This process ends when we get a zero remainder, say $r_{n-1} = r_n q_{n+1} + 0$. Then r_n, the last nonzero remainder, is the greatest common divisor of a, b. Moreover, using these equations backwards, we may express r_n in the form of a linear combination $au + bv$ for some integers u, v.

Proof. The case $b \mid a$ is clear. For $b \nmid a$, the process ends since $r_0 > r_1 > r_2 \cdots \ge 0$. The fact that the last nonzero remainder r_n is the greatest common divisor follows from part 6 in Theorem 7.16. Putting r_n in the form $au + bv$ is a straightforward computation. $\qquad\square$

Example 7.18. *Let's find gcd(306, 657). We have*

$$657 = 306 \cdot 2 + 45$$
$$306 = 45 \cdot 6 + 36$$
$$45 = 36 \cdot 1 + 9$$
$$36 = 9 \cdot 4 + 0,$$

hence gcd(306, 657) = 9. Moreover,

$$9 = 45 - 36 = 45 - (306 - 45 \cdot 6) = 306 + 45 \cdot 7 =$$

$$= 306 + (657 - 306 \cdot 2) \cdot 7 = (-13) \cdot 306 + 7 \cdot 657.$$

Corollary 7.19. *The Euclidean algorithm may be used to find the greatest common divisor of any two nonzero integers a, b.*

Proof. Indeed, we may reduce to the case of positive integers since

$$\gcd(a, b) = \gcd(a, -b) = \gcd(-a, b) = \gcd(-a, -b).$$

$\square$

Definition 7.20. *An integer p is prime if and only if p has exactly four divisors.*

For example, $\pm 2, \pm 13, \pm 19$ are primes; $0, \pm 1, \pm 6$ are not. An integer other than 0 or ± 1 that is not prime is called a *composite*. For example, $45 = 3 \cdot 3 \cdot 5$ is composite.

Lemma 7.21. *If p is prime and $p \mid xy$, then $(p \mid x) \vee (p \mid y)$.*

Proof. Assume p is prime and $p \mid xy$. Let $d = \gcd(p, x)$. Since p is prime, there are two possibilities: $d = |p|$ or $d = 1$. In the first case $p \mid x$. In the second case, by Lemma 7.12, $p \mid y$. $\square$

Corollary 7.22. *If p is a prime and $p \mid x_1 x_2 \cdots x_n$, then there is j such that $p \mid x_j$.*

Theorem 7.23. *(Fundamental theorem of arithmetic) Each integer except 0 and ± 1 is a prime, or it can be written in a unique way as a product of primes if we disregard the $\pm$ signs and the order of factors.*

Proof. We proved part of this theorem by complete induction for positive integers, see Theorem 6.37 in the previous chapter. If n is negative, then $-n$ is prime or $-n = p_1 p_2 \cdots p_k$ with p_j primes $j = 1, ..., k$. Then n is prime or $n = (-p_1) p_2 \cdots p_k$.

For the uniqueness part, assume that $n = p_1 p_2 \cdots p_k = q_1 q_2 \cdots q_m$ with p_i, q_j primes for $i = 1, ..., k, j = 1, ..., m$. We want to show that $k = m$ and that, after reordering and relabeling if necessary, $p_1 = \pm q_1, p_2 = \pm q_2, ..., p_k =$

$\pm q_k$. We have that $p_1 \mid q_1 q_2 \cdots q_m$. By Corollary 7.22, p_1 must divide one of the q_j. By reordering and relabeling, we may assume $p_1 \mid q_1$. It follows that $p_1 = \pm q_1$. Canceling p_1, we get $p_2 p_3 \cdots p_k = \pm q_2 q_3 \cdots q_m$ (assuming $k \geq 2$), hence $p_2 \mid q_2 q_3 \cdots q_m$. Repeating the argument, we eliminate one prime at a time on each side. If $k < m$, then after k steps, we get $1 = \pm q_{k+1} \cdots q_m$, which is impossible, since q_j are primes. A similar argument shows that $k > m$ does not work either. It remains that $k = m$ and $p_i = \pm q_i$ for all $i = 1, ..., k$. $\quad\square$

Corollary 7.24. *Consider two integers a, b with $|a|, |b| \geq 2$ and decompose them into a product of primes. Then $\gcd(a, b)$ is the absolute value of the product of the common primes from the decomposition with the least exponent, and $\mathrm{lcm}(a, b)$ is the absolute value of the product of all primes from the decomposition with the largest exponent.*

Example 7.25. *Let $a = -10140, b = 2600$. We have*

$$-10140 = -2^2 \cdot 3 \cdot 5 \cdot 13^2 \quad and \quad 2600 = 2^3 \cdot 5^2 \cdot 13.$$

Then $\gcd(-10140, 2600) = 2^2 \cdot 5 \cdot 13 = 260$ and $\mathrm{lcm}(-10140, 2600) = 2^3 \cdot 3 \cdot 5^2 \cdot 13^2 = 101400$.

Example 7.26. *Find all integer solutions to the (Diophantine) equations*

$$a)\ x^2 = y^3, \quad b)\ x^2 = y^4 - 77.$$

Solution. a) Notice that $y \geq 0$ since $x^2 \geq 0$. The obvious solutions are $x = y = 0$, $x = y = 1$, $x = -1, y = 1$. To find all solutions, assume $y \geq 2$, therefore $|x| \geq 2$. Since x and y have the same prime factors and in x^2 each prime has an exponent that is a multiple of 2 and in y^3 each prime has an exponent that is a multiple of 3, it follows that each exponent is a multiple of 6. Hence $x^2 = n^6 = y^3$ and all solutions are of the form $x = \pm n^3, y = n^2$ for $n \in \mathbb{Z}$.

b) The equation is equivalent to $y^4 - x^2 = 77$ or $(y^2 - x)(y^2 + x) = 77$. Since the only divisors of 77 are $\pm 1, \pm 7, \pm 11$ and ± 77 there are the following possibilities:
1) $y^2 - x = 1, y^2 + x = 77$, which gives $y^2 = 39$, no solution;
2) $y^2 - x = -1, y^2 + x = -77$, no solution;
3) $y^2 - x = 7, y^2 + x = 11$, which gives $y = \pm 3, x = 2$;
4) $y^2 - x = -7, y^2 + x = -11$, no solution;
5) $y^2 - x = 11, y^2 + x = 7$, which gives $y = \pm 3, x = -2$;
6) $y^2 - x = -11, y^2 + x = -7$, no solution;
7) $y^2 - x = 77, y^2 + x = 1$, no solution;
8) $y^2 - x = -77, y^2 + x = -1$, no solution.
The only solutions are $x = \pm 2, y = \pm 3$.

Example 7.27. *For the following congruence equations, find a solution $x \in \mathbb{Z}$ or show that no solution exists:*
1. $99x \equiv 18 \ (mod\ 30)$.
2. $91x \equiv 84 \ (mod\ 143)$.

Solution. 1. We must have $30 \mid 99x - 18 = 9(11x - 2)$. This implies $10 \mid 11x - 2$, so $11x = 10k + 2$. For example, we can take $k = 2$ and $x = 2$ is a solution.

2. We must have $143 = 11 \cdot 13 \mid 7(13x - 12)$. Since $13 \nmid 13x - 12$ because $13 \nmid 12$, there is no solution.

Example 7.28. *Prove that for all $n \geq 0$, $5^{2n} - 3^n$ is divisible by 11.*

Proof. This is true for $n = 0$ since $11 \mid 0$. Assume $11 \mid 5^{2k} - 3^k$ for some $k \geq 0$ and let's prove that $11 \mid 5^{2k+2} - 3^{k+1}$. We have

$$5^{2k+2} - 3^{k+1} = 25 \cdot 5^{2k} - 3 \cdot 3^k = 25(5^{2k} - 3^k) + 22 \cdot 3^k.$$

Since $11 \mid 5^{2k} - 3^k$ and $11 \mid 22$ it follows that $11 \mid 5^{2k+2} - 3^{k+1}$. $\qquad\square$

7.3 Integers in base 10 and divisibility tests

We are used to writing numbers in base 10. When we write 4712, we mean $4 \cdot 10^3 + 7 \cdot 10^2 + 1 \cdot 10^1 + 2 \cdot 10^0$. Any positive integer n has a representation in base 10

$$n = a_k a_{k-1} \cdots a_0 = a_k \cdot 10^k + a_{k-1} \cdot 10^{k-1} + \cdots + a_1 \cdot 10 + a_0,$$

where each a_i is a digit from 0 to 9 and we omit the leading zeros. This representation extends easily to all integers, by adding the number 0 and by using a minus sign for negative integers.

Let n be a positive integer. Recall that two integers a, b are congruent modulo n if n divides $a - b$. We write $a \equiv b (mod\ n)$. We proved in Example 5.27 that this is an equivalence relation on $\mathbb{Z}$.

Theorem 7.29. *If $a \equiv b(mod\ n)$ and $c \equiv d(mod\ n)$, then $a + c \equiv b + d(mod\ n)$ and $ac \equiv bd(mod\ n)$.*

Proof. Since $n \mid b - a$ and $n \mid d - c$ we get $n \mid (b - a) + (d - c) = b + d - (a + c)$, hence $a + c \equiv b + d(mod\ n)$. Now $bd - ac = bd - ad + ad - ac = (b - a)d + a(d - c)$ and it follows that $n \mid bd - ac$. $\qquad\square$

Theorem 7.30. *Every positive integer $a_k a_{k-1} \cdots a_0$ written in base 10 is congruent modulo 9 to the sum of its digits $a_k + \cdots + a_0$.*

Proof. Since $10 \equiv 1(mod\ 9)$, we get $10^i \equiv 1(mod\ 9)$, hence $a_i \cdot 10^i \equiv a_i(mod\ 9)$ for all $0 \leq i \leq k$. We get $\displaystyle\sum_{i=0}^{k} a_i \cdot 10^i \equiv \sum_{i=0}^{k} a_i(mod\ 9)$. $\qquad\square$

Corollary 7.31. *An integer is a multiple of 9 iff the sum of its digits is a multiple of 9.*

Theorem 7.32. *An integer with representation $a_k a_{k-1} \cdots a_0$ in base 10 is divisible by 11 iff the alternating sum of the digits $\sum_{i=0}^{k} (-1)^i a_i$ is divisible by 11.*

Proof. We use the congruence $10 \equiv -1 (\text{mod } 11)$ to get $a_i \cdot 10^i \equiv (-1)^i a_i (\text{mod } 11)$ and $\sum_{i=0}^{k} a_i \cdot 10^i \equiv \sum_{i=0}^{k} (-1)^i a_i (\text{mod } 11)$. $\square$

Theorem 7.33. *(Divisibility tests by $2, 4, 5, 6, 8, 10$). Let $n = a_k a_{k-1} \cdots a_0$ in base 10. Then*

a) n is divisible by 2 iff the last digit a_0 is even;

b) n is divisible by 4 iff the number formed by the last two digits $a_1 a_0$ is divisible by 4;

c) n is divisible by 5 iff the last digit a_0 is 0 or 5;

d) n is divisible by 6 iff the last digit a_0 is even and the sum of the digits is divisible by 3;

e) n is divisible by 8 iff the number formed by the last three digits $a_2 a_1 a_0$ is divisible by 8; and

f) n is divisible by 10 iff the last digit a_0 is 0.

Proof. To prove b), we write $n = a_k a_{k-1} \cdots a_2 \cdot 10^2 + a_1 a_0$ and $4 \mid 100$. To prove e), we write $n = a_k a_{k-1} \cdots a_3 \cdot 10^3 + a_2 a_1 a_0$ and $8 \mid 1000$. We leave the proof of the other divisibility tests as exercises. $\square$

There are different ways of representing integers, using a base other than 10. For example, in base 2, we use only the digits 0 and 1 and the number 7 is binary represented as 111. An integer with binary representation $a_k a_{k-1} \cdots a_1 a_0$ is divisible by 2 iff the last digit a_0 is 0. It is divisible by 4 iff the last two digits a_1, a_0 are 0.

7.4 Exercises

Exercise 7.34. *Prove that*

a. $|x| = \max(x, -x)$.

b. $\big||x| - |y|\big| \leq |x - y|$.

c. $a > 0 \Rightarrow (|x| > a \Leftrightarrow (x > a) \vee (x < -a))$.

Exercise 7.35. *Suppose it is now 8 a.m. in Boston. What time will be in 6538 hours?*

Exercise 7.36. *Prove that for nonzero integers x, y, z, t we have:*

a) 1 is a common divisor of x and 1.

b) x is a common divisor of x and x.

c) 1 is a common divisor of x and y.

d) If z is a common divisor of x and y and t | z, then t is a common divisor of x and y.

e) If z is a common divisor of x and y and t | z, then z/t is a common divisor of x/t and y/t.

Exercise 7.37. *Prove that for nonzero integers x, y, z, t we have:*

a) x is a common multiple of x and 1.

b) x is a common multiple of x and x.

c) x · y is a common multiple of x and y.

d) If z is a common multiple of x and y and z | t, then t is a common multiple of x and y.

e) If z is a common multiple of x and y, t | x, t | y, and t | z, then z/t is a common multiple of x/t and y/t.

Exercise 7.38. *Prove that for each integer x, there exists a prime p such that x < p.*

Exercise 7.39. *Find the greatest common divisor of the following pairs of numbers and express it as a linear combination au + bv:*

1. a = 56, b = 72.

2. a = 24, b = 138.

3. a = 143, b = 227.

4. a = 272, b = 1479.

Exercise 7.40. *Prove or disprove: If a | (b + c), then a | b or a | c.*

Exercise 7.41. *Prove that gcd(n, n + 1) = 1 for any integer n.*

Exercise 7.42. *What are the possible values for gcd(n, n + 6)?*

Exercise 7.43. *Use induction to show that if gcd(a, b) = 1, then gcd(a, b^n) = 1 for all n ≥ 1.*

Exercise 7.44. *For a, b, c ∈ ℤ \ {0} we defined gcd(a, b, c) to be the largest positive integer which divides a, b, c. Prove that there are integers s, t, u such that*

$$gcd(a, b, c) = sa + tb + uc.$$

Exercise 7.45. *Find the integer solutions of $x^4 = 4y^2 + 4y - 23$.*

Exercise 7.46. *Verify that $x^2 + x + 41$ is prime for all integers $-40 \leq x \leq 40$, but for x = 41 it is not prime.*

Exercise 7.47. *Show that an integer is a multiple of 3 iff the sum of its digits (in base 10) is a multiple of 3.*

Exercise 7.48. *Prove that if n is odd, then $n^2 \equiv 1 \pmod 8$.*

Exercise 7.49. *Prove that for any integer n we have $n^3 \equiv n(mod\,6)$.*

Exercise 7.50. *For the following congruence equations, find a solution $x \in \mathbb{Z}$ or show that no solution exists:*
   1. $x^2 + x + 1 \equiv 0 \,(mod\,7)$.
   2. $x^2 \equiv 2 \,(mod\,5)$.
   2. $x^2 + x + 1 \equiv 0 \,(mod\,5)$.

Exercise 7.51. *State some divisibility tests for positive integers represented in a base other than 10.*

Exercise 7.52. **Suppose n is a positive integer such that both $2n + 1$ and $3n + 1$ are perfect squares. Prove that n is divisible by 40.*

8

Cardinality: Finite sets, infinite sets

Learning how to count the elements of a set is a great achievement. Of course, this is easy for (small) finite sets like $\{a, b, c\}$ or $\{1, 3, 5, 7, 9, 11, 13\}$. What if the set is large or infinite? We will see that there are different flavors of infinity, something that puzzled people for many years. For example, there are infinitely many rationals and infinitely many irrationals. How do we compare the two sets? The correct way to compare infinite sets is a notion called *cardinality*, which is defined using bijective functions.

8.1 Equipotent sets

Definition 8.1. *Fix a universe U. Two sets $A, B \subseteq U$ have the same cardinality or are equipotent, written $A \approx B$, if there is a bijection $f : A \to B$.*

Example 8.2. *Let X be a set with ten elements, let S be the set of all six-element subsets of X, and let T be the set of all four-element subsets of X. Then $S \approx T$. Indeed, define $f : S \to T, f(A) = X \setminus A$. Then f is one-to-one and onto. Its inverse is $f^{-1} : T \to S, f^{-1}(B) = X \setminus B$. We will see later that both sets S and T have 210 elements.*

Example 8.3. *Let E denote the set of even integers, and define $f : \mathbb{Z} \to E, f(n) = 2n$. Then f is a bijection, hence $E \approx \mathbb{Z}$. This example illustrates the fact that $\mathbb{Z}$ is equipotent with a proper subset.*

Example 8.4. *Let $a, b \in \mathbb{R}$ with $a < b$. Then $f : (a, b) \to (0, 1), f(x) = \dfrac{x - a}{b - a}$ is a bijection and hence $(a, b) \approx (0, 1)$.*

Example 8.5. *The function $\arctan : \mathbb{R} \to (-\pi/2, \pi/2)$ is a bijection, so $\mathbb{R} \approx (-\pi/2, \pi/2)$.*

Theorem 8.6. *The relation $\approx$ is an equivalence relation on $\mathcal{P}(U)$, where U is a fixed universe. The equivalence class of A is denoted $|A|$ and is called the cardinality of A.*

Proof. Indeed, $\approx$ is reflexive, since for A not empty, $id_A : A \to A$ is a bijection. Note that, using the empty bijection, we also have $\emptyset \approx \emptyset$.

If $A \approx B$, let $f : A \to B$ be a bijection. Then $f^{-1} : B \to A$ is also a bijection, hence $B \approx A$, so $\approx$ is symmetric.

Assume $A \approx B$ and $B \approx C$. Then there are bijections $f : A \to B$ and $g : B \to C$. Consider $g \circ f : A \to C$. Then $g \circ f$ is a bijection, hence $A \approx C$, and $\approx$ is transitive. $\qquad\square$

Corollary 8.7. *We have* $|\mathbb{R}| = |(0, 1)|$.

Cardinals are generalizing natural numbers. Indeed, we can think of 0 as $|\emptyset|$, 1 as $|\{\emptyset\}|$, 2 as $|\{\emptyset, \{\emptyset\}\}|$, and in general $n + 1$ as $|n \cup \{n\}|$. The idea is that all sets containing exactly n elements have cardinality n.

Theorem 8.8. *Let A, B, C, D be sets such that $A \approx C$ and $B \approx D$. Then $A \times B \approx C \times D$.*

Proof. Let $f : A \to C$ and $g : B \to D$ be bijections. Define $f \times g : A \times B \to C \times D, (f \times g)(\langle a, b \rangle) = \langle f(a), g(b) \rangle$. Then $f \times g$ is a bijection. $\qquad\square$

Theorem 8.9. *Let A, B, C, D be sets such that $A \approx B, C \approx D, A \cap C = \emptyset$ and $B \cap D = \emptyset$. Then $A \cup C \approx B \cup D$.*

Proof. Consider bijections $f : A \to B$ and $g : C \to D$ and define

$$f \cup g : A \cup C \to B \cup D, (f \cup g)(x) = \begin{cases} f(x) & \text{if } x \in A \\ g(x) & \text{if } x \in C. \end{cases}$$

Then $f \cup g$ is well defined since $A \cap C = \emptyset$. It is one-to-one since f and g are one-to-one. It is onto since f, g are onto and $B \cap D = \emptyset$. $\qquad\square$

Definition 8.10. *For arbitrary sets A, B in a fixed universe U, we write $|A| \leq |B|$ if there is a one-to-one function $f : A \to B$ and $|A| < |B|$ if $|A| \leq |B|$ and $A \not\approx B$.*

Theorem 8.11. *(Cantor) If A is a set, then $|A| < |\mathcal{P}(A)|$.*

Proof. This is clear for $A = \emptyset$ because $|\emptyset| = 0 < 1 = |\{\emptyset\}|$. Assume $A \neq \emptyset$ and define $f : A \to \mathcal{P}(A)$ by $f(a) = \{a\}$. Then f is injective, since $f(a_1) = f(a_2)$ implies $\{a_1\} = \{a_2\}$, hence $a_1 = a_2$. We deduce that $|A| \leq |\mathcal{P}(A)|$. To prove the strict inequality, assume that there is a bijection $g : A \to \mathcal{P}(A)$, and define

$$B = \{a \in A : a \notin g(a)\}$$

(recall that $g(a) \subseteq A$). Then $B \subseteq A$, hence $B \in \mathcal{P}(A)$. Since g is onto, there is $b \in A$ with $g(b) = B$. Let's determine if $b \in B$ or not. If $b \in B$, then $b \notin g(b) = B$, a contradiction. If $b \notin B$, then $b \in g(b) = B$, a contradiction. It follows that g cannot be onto, hence $|A| < |\mathcal{P}(A)|$. Notice the similarity of this proof with Russell's paradox. $\qquad\square$

Theorem 8.12. * *(Cantor–Bernstein) If $|A| \leq |B|$ and $|B| \leq |A|$, then $|A| = |B|$.*

Proof. Since $|A| \leq |B|$ and $|B| \leq |A|$, there are $B^* \subseteq B$ with $A \approx B^*$ and $A^* \subseteq A$ with $B \approx A^*$. Consider the bijections $f : A \to B^*$ and $g : B \to A^*$ with inverses $f^{-1} : B^* \to A$, $g^{-1} : A^* \to B$. We will construct a bijection $h : A \to B$. Let $a \in A$. If $a \in A^*$, then $g^{-1}(a) \in B$. Let's call $g^{-1}(a)$ the first ancestor of a. If $g^{-1}(a) \in B^*$, then $f^{-1}(g^{-1}(a)) \in A$, which we call the second ancestor of a. If $f^{-1}(g^{-1}(a)) \in A^*$, then $g^{-1}(f^{-1}(g^{-1}(a))) \in B$, the third ancestor. We continue this process.

For each $a \in A$, one of the three possibilities holds:
1) a has infinitely many ancestors.
2) a has a last ancestor belonging to A.
3) a has a last ancestor belonging to B.
Define

$$A_\infty = \{a \in A : a \text{ has infinitely many ancestors}\}$$

$$A_0 = \{a \in A : a \text{ has an even number of ancestors}\}$$

$$A_1 = \{a \in A : a \text{ has an odd number of ancestors}\}.$$

Note that $A_\infty \subseteq A^*$, $A \setminus A^* \subseteq A_0$, and $A = A_\infty \cup A_0 \cup A_1$ with A_∞, A_0, A_1 mutually disjoint. In a similar way we decompose $B = B_\infty \cup B_0 \cup B_1$. We claim that f takes A_∞ onto B_∞, and takes A_0 onto B_1, while g^{-1} sends A_1 onto B_0. Indeed, if $a \in A$ has infinitely many ancestors, then $f(a) \in B$ has infinitely many ancestors; if $a \in A$ has an even number of ancestors, then $f(a) \in B$ has an odd number of ancestors, and if $a \in A$ has an odd number of ancestors, then $g^{-1}(a) \in B$ has an even number of ancestors. We can define

$$h : A \to B, h(x) = \begin{cases} f(x) & x \in A_\infty \cup A_0 \\ g^{-1}(x) & x \in A_1. \end{cases}$$

Since $f|_{A_\infty} : A_\infty \to B_\infty$, $f|_{A_0} : A_0 \to B_1$ and $g^{-1}|_{A_1} : A_1 \to B_0$ are bijections, we conclude that h is a bijection and $|A| = |B|$. $\qquad\square$

8.2 Finite and infinite sets

Definition 8.13. *Let $\mathbb{N}_0 = \emptyset$ and for $n \geq 1$, denote $\mathbb{N}_n = \{0, 1, 2, ..., n-1\}$. A set A is finite iff $A \approx \mathbb{N}_n$ for some $n \in \mathbb{N}$. In this case we say that A has n elements and write $|A| = n$. (Note in particular that $\emptyset$ is finite and it has 0 elements). A set is infinite if it is not finite. The sets $\mathbb{P}, \mathbb{N}, \mathbb{Z}, \mathbb{Q}, \mathbb{R}, \mathbb{R} \setminus \mathbb{Q}, \mathbb{C}$ are infinite. Also, the set of integer primes is infinite.*

Theorem 8.14. *If A, B are finite and disjoint, then $A \cup B$ is finite and $|A \cup B| = |A| + |B|$.*

Proof. Let $|A| = n, |B| = m$. We can find a bijection $f : A \to \{1, 2, ..., n\}$ and a bijection $g : B \to \{n+1, n+2, ..., n+m\}$. Then

$$h : A \cup B \to \{1, 2, ..., n+m\}, h(x) = \begin{cases} f(x) & \text{if} \quad x \in A \\ g(x) & \text{if} \quad x \in B \end{cases}$$

is a well-defined bijection and shows that $|A \cup B| = n + m$. $\qquad\square$

Recall that the disjoint union of two sets A, B is defined as

$$A \sqcup B = A \times \{1\} \cup B \times \{2\}.$$

For A, B finite, it follows that $|A \sqcup B| = |A| + |B|$.

Theorem 8.15. *If $A \subset B$ and B is finite nonempty, then A is finite and $|A| < |B|$.*

Proof. Let $|B| = n$. If $n = 1$, then $A = \emptyset$ and $|A| = 0 < 1$. Suppose that the result is true for $n = k$. Consider $A \subset B$ with $|B| = k+1$. We can assume $B = \{b_1, b_2, ..., b_k, b_{k+1}\}$. There are two cases: $b_{k+1} \notin A$ and $b_{k+1} \in A$. In the first case, $A \subset \{b_1, b_2, ...b_k\}$ and we are done. In the second case, it follows that $A_1 = A \cap \{b_1, b_2, ...b_k\} \subset \{b_1, b_2, ...b_k\}$ and $A = \{b_{k+1}\} \cup A_1$. We know that $|A_1| < k$ and therefore $|A| = 1 + |A_1| < k + 1$. $\qquad\square$

Theorem 8.16. *Every set containing an infinite subset is infinite. An infinite set is equipotent to a proper subset.*

Proof. If $A \subseteq B$ and A is infinite, then B is infinite. Indeed, we already proved the contrapositive: if B is finite and $A \subseteq B$, then A is finite.

For the second part, assume A is infinite. First we use the axiom of choice to prove that it contains an infinite set of the form $C = \{x_1, x_2, x_3, ...\}$. Since A is infinite, it is nonempty. Choose $x_1 \in A$. The set $A \backslash \{x_1\}$ is also infinite; choose $x_2 \in A \backslash \{x_1\}$. Inductively we can choose $x_n \in A \backslash \{x_1, x_2, ..., x_{n-1}\}$ for all $n \geq 2$. At each step, the set $A \backslash \{x_1, x_2, ..., x_{n-1}\}$ is not empty since A is infinite. Define $f : A \to A$ by $f(x_i) = x_{2i}$ for all $x_i \in C$ and $f(a) = a$ for all $a \notin C$. Then f is a bijection of A onto the proper subset $A_1 = A \backslash \{x_1, x_3, x_5, ...\}$. $\qquad\square$

In fact, if a set A is equipotent to a proper subset, then A cannot be finite. This property is taken sometimes as the definition of an infinite set.

8.3   Countable and uncountable sets

Definition 8.17. *A set A is called* countable *iff $A \approx \mathbb{N}$. A set is called* at most countable *if it is finite or countable. A set which is not at most countable is called* uncountable. *The cardinality of $\mathbb{N}$ is denoted by $\aleph_0$ (read: aleph zero), so $|\mathbb{N}| = \aleph_0$.*

It should be mentioned that some authors call a set A countable if A is finite or $A \approx \mathbb{N}$.

Examples 8.18. *1. The set $\mathbb{P}$ of positive integers is countable, since $f : \mathbb{P} \to \mathbb{N}, f(n) = n - 1$ is a bijection.*

2. The set $\mathbb{Z}$ is countable. Indeed, the function

$$f : \mathbb{N} \to \mathbb{Z}, f(n) = \begin{cases} k & \text{if } n = 2k \\ -k & \text{if } n = 2k + 1 \end{cases}$$

is a bijection (exercise!).

3. The set E of even integers is countable. Indeed, $f : \mathbb{Z} \to E, f(k) = 2k$ is a bijection.

Theorem 8.19. *Let A be a nonempty set. The following are equivalent:*

(1) There is an onto function $f : \mathbb{N} \to A$.

(2) There is one-to-one function $g : A \to \mathbb{N}$.

(3) A is at most countable.

Proof. $(1) \Rightarrow (2)$. Let $f : \mathbb{N} \to A$ be a surjection. Define $g : A \to \mathbb{N}$ by $g(a) =$ the least element of $f^{-1}(a)$. The function g is well defined because f is onto, so $f^{-1}(a) \neq \emptyset$. To show that g is one-to-one, notice that for $a_1 \neq a_2$, the sets $f^{-1}(a_1)$ and $f^{-1}(a_2)$ are disjoint, so $g(a_1) \neq g(a_2)$.

$(2) \Rightarrow (3)$. Fix $g : A \to \mathbb{N}$ one-to-one. To prove that A is at most countable, notice that it suffices to show that any subset B of $\mathbb{N}$ is at most countable, since A is in bijection with $g(A)$. If B is finite, it is at most countable by definition. Assume $B \subseteq \mathbb{N}$ is infinite, and let's construct a bijection $h : \mathbb{N} \to B$. Let $h(0)$ be the least element of B, let $h(1)$ be the least element of $B \setminus \{h(0)\}$, and in general let $h(n)$ be the least element of $B \setminus \{h(0), ..., h(n-1)\}$. For all n, the set $B \setminus \{h(0), ..., h(n-1)\}$ is not empty since B is infinite, and the least element exists by the well-ordering principle. Notice that by construction, h is one-to-one since for $m < n$ the element $h(m)$ belongs to $\{h(0), ..., h(n-1)\}$, so $h(m) \neq h(n)$. In particular $h(\mathbb{N})$ is infinite. To show that h is onto, let $b \in B$ and choose $n \in \mathbb{N}$ such that $h(n) > b$. Let m be the smallest natural number such that $h(m) \geq b$. Then for all $j < m$ we must have $h(j) < b$, so $b \notin h(\{0, 1, ..., m-1\})$. By definition, $h(m)$ is the smallest element of $B \setminus h(\{0, 1, ..., m-1\})$, so $h(m) \leq b$. It follows that $h(m) = b$ and h is onto.

$(3) \Rightarrow (1)$. Suppose A is at most countable. If A is infinite, there is a bijection $f : \mathbb{N} \to A$ by definition; in particular this f is onto. If A is finite, we can find a bijection $f : \mathbb{N}_n \to A$ for some $n \geq 1$. We can extend f to a surjection $\tilde{f} : \mathbb{N} \to A$ by defining $\tilde{f}(m) = f(0)$ for $m \geq n$. $\square$

Corollary 8.20. *A subset of a countable set is at most countable.*

Theorem 8.21. *If A is finite and B is at most countable, then $A \cup B$ is at most countable.*

Proof. If B is finite, then $A \cup B$ is finite, hence at most countable. Assume B is infinite. Since $A \cup B = (A \setminus B) \cup B$ and $A \setminus B$ is finite, it suffices to consider disjoint sets. Consider bijections $f : A \to \mathbb{N}_n$ for some $n \geq 0$ and $g : B \to \mathbb{N}$. Define $h : A \cup B \to \mathbb{N}, h(a) = f(a)$ if $a \in A$ and $h(b) = n + g(b)$ if $b \in B$. Since $A \cap B = \emptyset$, h is well defined, and it is one-to-one and onto, since f and g are also. $\qquad\square$

Theorem 8.22. *If A, B are at most countable sets, then $A \cup B$ is at most countable.*

Proof. As before, we may assume that A, B are disjoint and infinite. To show that $A \cup B$ is countable, we first define bijections $f : A \to E$ and $g : B \to \mathbb{Z} \setminus E$, where E is the set of even integers. Define $h : A \cup B \to \mathbb{Z}$ as $h = f \cup g$. Then h is a bijection. $\qquad\square$

Corollary 8.23. *A finite union of at most countable sets is at most countable.*

Theorem 8.24. *The set $\mathbb{N} \times \mathbb{N}$ is countable.*

Proof. Let $f : \mathbb{N} \times \mathbb{N} \to \mathbb{P}, f(m, n) = 2^m(2n + 1)$. Then f is a bijection. Indeed, if $f(m, n) = f(m', n')$, then $2^{m'}(2n' + 1) = 2^m(2n + 1)$, hence $m = m'$ and $n' = n$ and f is one-to-one. Given a positive integer k, then write $k = 2^m(2n + 1)$ for some $m \geq 0$ and $n \geq 0$, hence f is onto. Since $\mathbb{P} \approx \mathbb{N}$, we get that $\mathbb{N} \times \mathbb{N} \approx \mathbb{N}$. $\qquad\square$

Theorem 8.25. *The set $\mathbb{Q}^+$ of positive rational numbers is countable.*

Proof. Write

$$\mathbb{Q}^+ = \left\{ \frac{p}{q} : p, q \in \mathbb{P} \text{ and } \gcd(p, q) = 1 \right\}$$

and define $f : \mathbb{Q}^+ \to \mathbb{N} \times \mathbb{N}, f(p/q) = \langle p, q \rangle$. Then f is one-to-one, hence $|\mathbb{Q}^+| \leq \aleph_0$. Since $g : \mathbb{P} \to \mathbb{Q}^+, g(p) = \dfrac{p}{1}$ is also one-to-one, we get $\aleph_0 \leq |\mathbb{Q}^+|$, hence equality. $\qquad\square$

Corollary 8.26. *The set of rational numbers $\mathbb{Q}$ is countable.*

Proof. If $\mathbb{Q}^-$ denotes the set of negative rational numbers, then $f : \mathbb{Q}^+ \to \mathbb{Q}^-, f(x) = -x$ is a bijection. Now use the fact that $\mathbb{Q} = \mathbb{Q}^+ \cup \{0\} \cup \mathbb{Q}^-$. $\quad\square$

Theorem 8.27. *The interval $(0, 1)$ is uncountable.*

Proof. Suppose $f : \mathbb{P} \to (0, 1)$ is a bijection. We list all real numbers in $(0, 1)$ in decimal form, not ending with an infinite string of nines:

$$f(1) = 0.a_{11}a_{12}a_{13}...$$

$$f(2) = 0.a_{21}a_{22}a_{23}...$$

$$f(3) = 0.a_{31}a_{32}a_{33}...$$

$$\vdots$$

$$f(n) = 0.a_{n1}a_{n2}a_{n3}...$$

$$\vdots$$

Define

$$b_k = \begin{cases} 2, & \text{if } a_{kk} \neq 2 \\ 4, & \text{if } a_{kk} = 2. \end{cases}$$

Notice that the number $b = 0.b_1b_2b_3... \in (0,1)$ does not appear on the list because $b_k \neq a_{kk}$ and $b_k \neq 9$. This is a contradiction with the fact that f is a bijection, so $(0,1)$ is uncountable. This proof technique is called Cantor's diagonal argument. $\qquad\square$

Definition 8.28. *We define the sum of cardinals as* $|A| + |B| = |A \sqcup B|$ *and the product of cardinals as* $|A| \cdot |B| = |A \times B|$.

In particular, $\aleph_0 + \aleph_0 = \aleph_0$ and $\aleph_0 \cdot \aleph_0 = \aleph_0$.

We have $\aleph_0 < |\mathcal{P}(\mathbb{N})| = 2^{\aleph_0}$, where for a set A we define $2^{|A|}$ as $|\mathcal{P}(A)|$. The cardinality of $(0,1)$ or $\mathbb{R}$ is denoted by $\mathbf{c}$, called the *continuum*.

Most people accept the *continuum hypothesis*: there is no set X with $\aleph_0 < |X| < \mathbf{c}$. This hypothesis appeared in the work of Georg Cantor in 1878 and it is independent of the other axioms for set theory, as proved by Paul Cohen in 1963.

Corollary 8.29. *Assuming the continuum hypothesis, we have* $\mathbf{c} = 2^{\aleph_0}$.

8.4 Exercises

Exercise 8.30. *Let A be a finite set and let $f : A \to A$ be a function. Prove that f is onto iff f is one-to-one.*

Exercise 8.31. *If A is an infinite set, construct $f : A \to A$, which is one-to-one but not onto.*

Exercise 8.32. *Let A be a set. Recall that $\{0,1\}^A$ denotes the set of all functions $f : A \to \{0,1\}$. Using characteristic functions, prove that $|\{0,1\}^A| = |\mathcal{P}(A)|$.*

Exercise 8.33. *Consider a bijection $\phi : \mathbb{Q} \to \mathbb{N}$ and define a relation R on $\mathbb{Q}$ such that xRy iff $\phi(x) \leq \phi(y)$. Prove that $(\mathbb{Q}, R)$ is a well-ordered set.*

Exercise 8.34. *Show that a countable union of countable sets is countable.*

Exercise 8.35. *Denote by $\mathcal{P}_f(\mathbb{N})$ the set of finite subsets of $\mathbb{N}$. Prove that $\mathcal{P}_f(\mathbb{N})$ is countable.*

Exercise 8.36. *Prove that the set of irrational numbers is uncountable. Deduce that there are more irrational numbers than rationals.*

Exercise 8.37. *Let $S = \{0, 1\}^{\mathbb{N}}$ be the set of all infinite sequences of 0s and 1s. Use Cantor's diagonal argument to prove that S is uncountable.*

Exercise 8.38. *Find a bijection $f : (0, 1) \to [0, 1]$. (Hint: Choose a countable subset $\{x_0, x_1, x_2, ...\}$ of $(0, 1)$ and define $f(x_0) = 0, f(x_1) = 1$ and for $n \geq 2$ let $f(x_n) = x_{n-2}$. Extend f to a bijection).*

Exercise 8.39. *Suppose there are injective functions $f : A \to B, g : B \to C$ and $h : C \to A$. Prove that $A \approx B \approx C$.*

9

Counting techniques and combinatorics

In this chapter we will learn how to count the number of elements in certain finite sets, using the inclusion-exclusion principle, the multiplication principle, and more. We enrich our methods of proofs, by using the pigeonhole principle, by using parity, and by using other techniques.

We will introduce permutations and combinations, binomial coefficients, recursive sequences, and recurrence relations.

9.1 Counting principles

Theorem 9.1. *Let* $A_1, ..., A_n$ *be disjoint finite sets. Then*

$$|A_1 \cup \cdots \cup A_n| = |A_1| + \cdots + |A_n|.$$

Proof. We have seen in Theorem 8.14 that for A, B disjoint we have $|A \cup B| = |A| + |B|$. The proof now proceeds by induction. $\square$

Example 9.2. *Let A be the set of all integers n from 1 to 100 which have at least one digit of 4. Then $|A| = 19$. Indeed, we have $A = A_1 \cup ... \cup A_{10}$, where $A_1 = \{4\}, A_2 = \{14\}, A_3 = \{24\}, ..., A_5 = \{41, 42, ..., 49\}, ..., A_{10} = \{94\}$. Then $|A_5| = 10$ and $|A_i| = 1$ for $1 \le i \le 10$, $i \ne 5$. Thus $|A| = |A_1| + ... + |A_{10}| = 19$.*

Theorem 9.3. *Let A, B be finite sets. Then*

$$|A \cup B| = |A| + |B| - |A \cap B|.$$

Proof. We can write $A = (A \setminus B) \cup A \cap B$ and $B = (B \setminus A) \cup A \cap B$ and the sets $A \setminus B, B \setminus A$ and $A \cap B$ are disjoint. It follows that

$$|A \cup B| = |A \setminus B| + |B \setminus A| + |A \cap B| =$$

$$= |A| - |A \cap B| + |B| - |A \cap B| + |A \cap B| = |A| + |B| - |A \cap B|.$$

$\square$

Notice that by adding $|A|$ and $|B|$ we included the common elements of A and B twice, and this is the reason to subtract $|A \cap B|$, to exclude them once. This idea is generalized in the next principle of counting, where the signs are alternating.

Corollary 9.4. *(Inclusion-exclusion principle) Let $A_1, A_2, ..., A_n$ be finite sets. Then*

$$|A_1 \cup A_2 \cup \cdots \cup A_n| =$$

$$= \sum_{i=1}^{n} |A_i| - \sum_{i<j\leq n} |A_i \cap A_j| + \sum_{i<j<k\leq n} |A_i \cap A_j \cap A_k| - \cdots + (-1)^{n+1}|A_1 \cap A_2 \cap \cdots \cap A_n|.$$

Proof. We proceed by induction. The result is true for $n = 2$. Assuming the formula for k finite sets, let's prove it for $k+1$ finite sets $A_1, A_2, ..., A_k, A_{k+1}$. We have

$$|A_1 \cup A_2 \cup \cdots \cup A_k \cup A_{k+1}| =$$

$$= |A_1 \cup A_2 \cup \cdots \cup A_k| + |A_{k+1}| - |(A_1 \cup A_2 \cup \cdots \cup A_k) \cap A_{k+1}| =$$

$$= \sum_{i=1}^{k} |A_i| - \sum_{i<j\leq k} |A_i \cap A_j| + \sum_{i<j<l\leq k} |A_i \cap A_j \cap A_l| - \cdots + (-1)^{k+1}|A_1 \cap A_2 \cap \cdots \cap A_k| +$$

$$+ |A_{k+1}| - |(A_1 \cap A_{k+1}) \cup (A_2 \cap A_{k+1}) \cup \cdots \cup (A_k \cap A_{k+1})| =$$

$$= \sum_{i=1}^{k+1} |A_i| - \sum_{i<j\leq k+1} |A_i \cap A_j| + \sum_{i<j<l\leq k+1} |A_i \cap A_j \cap A_l| - \cdots$$

$$\cdots + (-1)^{k+2}|A_1 \cap A_2 \cap \cdots \cap A_k \cap A_{k+1}|.$$

$\square$

Example 9.5. *How many positive integers from 1 to 100 are divisible by 2, 3, or 5?*

Solution. There are 50 integers divisible by 2, 33 divisible by 3, 20 divisible by 5, 16 divisible by 6, 10 divisible by 10, 6 divisible by 15, and 3 divisible by 30. Using the inclusion-exclusion principle, the number of integers from 1 to 100 divisible by 2, 3, or 5 is

$$50 + 33 + 20 - 16 - 10 - 6 + 3 = 74.$$

Theorem 9.6. *(Multiplication principle) For $A_1, A_2, ..., A_n$ finite, we have*

$$|A_1 \times \cdots \times A_n| = |A_1| \cdot \cdots \cdot |A_n|.$$

Proof. Indeed, $|A \times B| = |A| \cdot |B|$ and we may use induction on n. $\square$

Theorem 9.7. *(Exponential principle) Recall that A^B denotes the set of functions $f : B \to A$. For A, B finite sets we have $|A^B| = |A|^{|B|}$.*

Proof. We use induction on $|B|$. For $|B| = 1$, the set A^B has exactly $|A|$ elements. Assume that the formula is true for $|B| = k$. Then for $|B| = k + 1$, let's assume $B = \{b_1, b_2, ..., b_k, b_{k+1}\}$. A function $f : B \to A$ is determined by $f \mid_{\{b_1, b_2, ..., b_k\}}$ and $f(b_{k+1})$. Since there are $|A|^k$ possibilities for the restriction $f \mid_{\{b_1, b_2, ..., b_k\}}$ and $|A|$ possibilities for $f(b_{k+1})$, we obtain that $|A^B| = |A|^k \cdot |A| = |A|^{k+1} = |A|^{|B|}$. $\square$

For infinite sets A, B, we define $|A|^{|B|}$ to be $|A^B|$. This is consistent with the exponential principle.

9.2 Pigeonhole principle and parity

Theorem 9.8. *(Pigeonhole principle) If $n + 1$ marbles are put into n boxes, then at least one box has more than one marble.*

This simple combinatorial principle was used by Dirichlet in number theory. You may replace the marbles by any other objects. This is also called the box principle, and it has many applications. We list some easy consequences:

1. Among 13 persons, there are two born in the same month.

2. If $qs + 1$ marbles are put into s boxes, then at least one box has more than q marbles.

3. A line ℓ in the plane of the triangle ABC does not pass through any of the vertices. Then ℓ cannot cut all sides of the triangle.

Here are some more involved examples.

Example 9.9. *Choose $n + 1$ numbers from the set $\{1, 2, ..., 2n\}$ Then one of these $n + 1$ numbers is divisible by another from this group.*

Proof. Denote the chosen numbers by $a_1, a_2, ..., a_{n+1}$ and write $a_i = 2^{k_i} b_i$ with $k_i \geq 0$ and b_i odd. We get $n + 1$ odd numbers $b_1, b_2, ..., b_{n+1}$ from the set $\{1, 3, ..., 2n - 1\}$, which has only n elements. It must be that $b_p = b_q$ for some indices p, q. Assuming $k_p \leq k_q$, we get $a_p \mid a_q$. $\square$

Example 9.10. *Let $a_1, a_2, ..., a_n$ be n integers, not necessarily distinct. Then there is a subset of these integers with a sum divisible by n.*

Proof. Let $s_1 = a_1, s_2 = a_1 + a_2, ..., s_n = a_1 + a_2 + \cdots + u_n$. If any of these sums is divisible by n, we are done. Otherwise, consider the remainders r_i when s_i is divided by n for each $i = 1, 2, ..., n$. Since there are only $n - 1$ possible nonzero remainders, two of the sums, say s_p and s_q, give the same remainders $r_s = r_q$. Then $n \mid s_p - s_q$. $\square$

Example 9.11. *Inside a room of area 5 you place 9 rugs, each of area 1. Then there are two rugs overlapping by at least 1/9 units of area.*

Proof. Suppose any two rugs overlap by less than 1/9. We place the rugs on the floor one by one. The first rug will cover an area of 1. The second rug will cover an area greater than 8/9, the third will cover an area that is more than 7/9, and so on. The last rug covers more than 1/9. But

$$1 + \frac{8}{9} + \frac{7}{9} + \cdots + \frac{1}{9} = \frac{9 + 8 + 7 + \cdots + 1}{9} = 5,$$

a contradiction.   □

Some problems involve parity of numbers. Recall that two integers have the same parity if they are both even or both odd.

Example 9.12. *Suppose we have two containers, one of 6 quarts and the other of 2 quarts. By filling them with water at the river and pouring one into the other, prove that we cannot measure 5 quarts of water.*

Proof. Note that our various filling and pouring operations correspond to adding and subtracting multiples of 2 and 6. Since all these numbers are even, we cannot obtain 5 as an answer. It follows that we cannot measure 5 quarts of water.   □

Example 9.13. *Consider a polyhedron with 13 vertices, and assume that each edge is assigned an electrical charge of $+1$ or -1. Prove that there must be a vertex such that the product of the charges of all the edges that meet at that vertex is $+1$.*

Proof. For each vertex v, consider the product P_v of the charges of the edges that meet at v. Since each edge is determined by two vertices, when we multiply all P_v's, we must get $+1$ because each edge appears twice. It follows that this product is $+1$ since we have an even number of -1. Since 13 is odd, we cannot have $P_v = -1$ for all v since $(-1)^{13} = -1$. It follows that there is a vertex v with $P_v = 1$.   □

9.3   Permutations and combinations

Definition 9.14. *A permutation of a set A is a bijection $\pi : A \to A$.*

For example, there are six possible "words" that can be formed with the letters A, B, C: $ABC, ACB, BAC, CBA, BCA, CAB$.

Theorem 9.15. *There are $n! = 1 \cdot 2 \cdot 3 \cdots n$ permutations of a set with $n \geq 1$ elements.*

Proof. Let $A = \{a_1, ..., a_n\}$ be a set with $n \geq 1$ elements. For a bijection $\pi : A \to A$, there are n possibilities for $\pi(a_1)$. Once we fix $\pi(a_1)$, there are $n-1$ possibilities for $\pi(a_2)$. Once we fix $\pi(a_2)$, there are $n-2$ possibilities for $\pi(a_3)$. Continuing in this way, there is only one possibility for $\pi(a_n)$. Multiplying, we get $n(n-1)\cdots 1 = n!$ bijections $\pi : A \to A$. $\qquad\square$

Definition 9.16. *A permutation of size k of n objects with $1 \leq k \leq n$ is an ordered list of length k of elements of a set A with n elements.*

For example, there are six "words" of two letters formed with the letters A, B, C: AB, BA, AC, CA, BC, CB.

Theorem 9.17. *If we denote by $P(n,k)$ the number of permutations of length k of a set with n elements, then*

$$P(n,k) = n(n-1)\cdots(n-k+1) = \frac{n!}{(n-k)!},$$

where by definition $0! = 1$.

Proof. Indeed, for the first position there are n choices, for the second position there are $n-1$ choices,..., for the k-th position there are $n-k+1$ choices, so

$$P(n,k) = n(n-1)\cdots(n-k+1).$$

We can multiply and divide by the product $(n-k)(n-k-1)\cdots 1$ to obtain the formula

$$P(n,k) = \frac{n!}{(n-k)!}.$$

$\qquad\square$

Definition 9.18. *Let A be a set with n elements, and let k be an integer such that $0 \leq k \leq n$. A subset of A with k elements is called a combination of size k chosen from A. The number of such combinations is denoted by $C(n,k)$. Another notation is C_n^k or $\binom{n}{k}$, which we read: n choose k.*

For example, the set $\{A, B, C\}$ has three subsets with two elements:

$$\{A, B\}, \{A, C\}, \{B, C\}.$$

Theorem 9.19. *For $n \geq 1$, we have*

$$C(n,k) = \frac{P(n,k)}{k!} = \frac{n!}{k!(n-k)!}.$$

Proof. Since we relax the condition that in the list of k elements the order counts, we get $P(n,k) = C(n,k) \cdot k!$, hence $C(n,k) = \dfrac{P(n,k)}{k!}$. Now we use the fact that $P(n,k) = \dfrac{n!}{(n-k)!}$. $\qquad\square$

Example 9.20. *From a club with 20 members, a president, a vice president, and a secretary are to be chosen. In how many ways can this be done?*

Solution. We have to choose three members of the club, where the order counts. The answer is

$$P(20,3) = 20 \cdot 19 \cdot 18 = 6840.$$

Example 9.21. *How many poker hands (5 cards) can you draw from a deck of 52 cards?*

Solution. There are $\dbinom{52}{5} = \dfrac{52 \cdot 51 \cdot 50 \cdot 49 \cdot 48}{1 \cdot 2 \cdot 3 \cdot 4 \cdot 5} = 2598960$ possible poker hands.

Theorem 9.22. *We have*

$$C(n,k) = C(n, n-k),$$

$$C(n+1, k) = C(n,k) + C(n, k-1)$$

and

$$(n-k)C(n,k) = nC(n-1,k).$$

Proof. Indeed,

$$\frac{n!}{k!(n-k)!} = \frac{n!}{(n-k)!(n-(n-k))!},$$

$$\frac{n!}{k!(n-k)!} + \frac{n!}{(k-1)!(n-k+1)!} = \frac{(n-k+1)n! + kn!}{k!(n+1-k)!} =$$

$$= \frac{(n+1)n!}{k!(n+1-k)!} = \frac{(n+1)!}{k!(n+1-k)!},$$

$$(n-k)\frac{n!}{k!(n-k)!} = \frac{n!}{k!(n-k-1)!} = n\frac{(n-1)!}{k!(n-1-k)!}.$$

$\square$

The numbers $\dbinom{n}{k}$ are also called binomial coefficients. They have a triangular representation, called *Pascal's triangle*.

$$
\begin{array}{ccccccccccccc}
& & & & & & 1 & & & & & & \\
& & & & & 1 & & 1 & & & & & \\
& & & & 1 & & 2 & & 1 & & & & \\
& & & 1 & & 3 & & 3 & & 1 & & & \\
& & 1 & & 4 & & 6 & & 4 & & 1 & & \\
& 1 & & 5 & & 10 & & 10 & & 5 & & 1 & \\
1 & & 6 & & 15 & & 20 & & 15 & & 6 & & 1 \\
\end{array}
$$

$\cdots$

In each row, we use the identity $C(n+1, k) = C(n, k) + C(n, k-1)$ to see that a binomial coefficient is the sum of the two adjacent coefficients in the line above (whenever this makes sense).

Theorem 9.23. *(The binomial formula) We have*

$$(x+y)^n = \binom{n}{0} x^n + \binom{n}{1} x^{n-1}y + \cdots + \binom{n}{n-1} xy^{n-1} + \binom{n}{n} y^n$$

$$= \sum_{m=0}^{n} \binom{n}{m} x^{n-m} y^m.$$

Proof. For $n = 1$ this is clear, since $(x+y)^1 = x + y = \binom{1}{0} x + \binom{1}{1} y.$

Assume

$$(x+y)^k = \sum_{m=0}^{k} \binom{k}{m} x^{k-m} y^m.$$

Then

$$(x+y)^{k+1} = (x+y)(x+y)^k =$$

$$= x^{k+1} + \sum_{m=1}^{k} \binom{k}{m} x^{k-m+1} y^m + \sum_{m=0}^{k-1} \binom{k}{m} x^{k-m} y^{m+1} + y^{k+1} =$$

$$= x^{k+1} + \sum_{m=1}^{k} \left[\binom{k}{m} + \binom{k}{m-1} \right] x^{k+1-m} y^m + y^{k+1} =$$

$$= \sum_{m=0}^{k+1} \binom{k+1}{m} x^{k+1-m} y^m.$$

$\square$

Corollary 9.24. *We have*

$$\sum_{k=0}^{n} \binom{n}{k} = 2^n. \tag{9.1}$$

$$\sum_{k=1}^{n} k \binom{n}{k} = n2^{n-1}. \tag{9.2}$$

$$\binom{2n}{n} = \sum_{k=0}^{n} \binom{n}{k}^2. \tag{9.3}$$

Proof. For (9.1), take $x = y = 1$ in the binomial formula.

For (9.2), differentiate $(1+x)^n = \sum_{k=0}^{n} \binom{n}{k} x^k$ and then take $x = 1$.

For (9.3), write $(1+x)^{2n} = (1+x)^n (1+x)^n$ and identify the coefficient of x^n both sides. $\square$

Exercise 9.25. *Show that*

1. $\binom{n}{2} + \binom{n+1}{2} = n^2$.

2. $\binom{n}{k} = \frac{n}{k}\binom{n-1}{k-1}$.

3. $\binom{a}{b}\binom{b}{c} = \binom{a}{a-c}\binom{a-c}{b-c}$ *for* $a \geq b \geq c \geq 0$.

Proof. 1. A direct computational proof is straightforward. A more interesting proof is to count the set of ordered pairs $\langle p, q \rangle$ with $p, q \in \{1, 2, ..., n\}$ in two different ways. If we consider the case $p < q$, then there are $\binom{n}{2}$ such pairs. For the case $p \geq q$, we can add 1 to the first component, and we count instead the ordered pairs $\langle p', q \rangle$ where $p' > q$ and $p', q \in \{1, 2, ..., n, n+1\}$. There are $\binom{n+1}{2}$ of them. Since the total of ordered pairs is n^2, we are done.

2. Again, a computational proof is easy. Here is a combinatorial interpretation: To choose a subset with k elements of $\{1, 2, ..., n\}$, we can pick an element first (there are n possibilities) and then we choose a subset with $k-1$ elements of the remaining $n-1$ elements (there are $\binom{n-1}{k-1}$ such subsets). We have to divide by k, since the first element in a subset with k elements can be any of its k elements.

3. Both sides count the number of ways to divide a set with a elements into three subsets with $a - b, b - c$, and c elements respectively. $\square$

9.4 Recursive sequences and recurrence relations

Definition 9.26. *Given a sequence* $s : \mathbb{N} \to \mathbb{R}$, *we write* $s = (s_n)_{n \geq 0}$, *where* $s_n = s(n)$. *The sequence* $(s_n)_{n \geq 0}$ *is called* recursive *if there is* $n_0 \in \mathbb{N}$ *such that for all* $n > n_0$, *the term* s_n *can be expressed as a function of* $s_0, s_1, ..., s_{n-1}$. *That function is called the* recurrence relation.

Example 9.27. *Let* $s_0 = 1$ *and let* $s_n = 2s_{n-1}$ *for* $n \geq 1$. *Then by induction we can prove that* $s_n = 2^n$.

Example 9.28. *Consider* n *straight lines in the plane such that no two are parallel and no three meet at a point (they are in general position). The number of regions* r_n *determined in the plane by the* n *lines satisfies* $r_0 = 1, r_{n+1} = r_n + n + 1$.

Example 9.29. *In a tennis tournament there are* $2n$ *participants. Show that the number* p_n *of pairings for the first round is* $p_n = \dfrac{(2n)!}{2^n \cdot n!}$.

Solution. For a fixed player, the partner can be chosen in $2n - 1$ ways and $(n - 1)$ pairs are left. Thus, $p_n = (2n - 1)p_{n-1}$. We get

$$p_n = (2n - 1)(2n - 3) \cdots 3 \cdot 1 = \frac{(2n)!}{2^n \cdot n!}.$$

Example 9.30. *The Fibonacci sequence* $(f_n)_{n \geq 0}$, *where*

$$f_0 = f_1 = 1, f_n = f_{n-1} + f_{n-2} \text{ for } n \geq 2$$

appears in nature and is related to the golden ratio $\dfrac{1 + \sqrt{5}}{2}$. *A formula for the general term is given below.*

We will use the following result, which reminds you about solving linear differential equations of second order:

Theorem 9.31. *Let* $(x_n)_{n \geq 0}$ *be a recursive sequence such that*

$$x_n = bx_{n-1} + cx_{n-2},$$

where $b, c \in \mathbb{R}$ *are fixed. If the characteristic equation*

$$t^2 - bt - c = 0$$

has distinct (complex) roots r_1, r_2, *then*

$$x_n = \alpha r_1^n + \beta r_2^n$$

for some α, β, *determined by* x_0, x_1.
    If the equation $t^2 - bt - c = 0$ *has repeated roots* $r_1 = r_2 = r$, *then* $x_n = \alpha r^n + \beta n r^n$ *for some* α, β.

For the Fibonacci sequence, the characteristic equation $t^2 - t - 1 = 0$ has roots
$$r_1 = (1 + \sqrt{5})/2, \quad r_2 = (1 - \sqrt{5})/2.$$

Since
$$\alpha + \beta = 1 \text{ and } \alpha r_1 + \beta r_2 = 1,$$

we get
$$\alpha = \frac{1 + \sqrt{5}}{2\sqrt{5}}, \quad \beta = \frac{\sqrt{5} - 1}{2\sqrt{5}},$$

hence
$$f_n = \frac{1}{\sqrt{5}} \left[\left(\frac{1 + \sqrt{5}}{2} \right)^{n+1} - \left(\frac{1 - \sqrt{5}}{2} \right)^{n+1} \right].$$

Definition 9.32. *(Generating functions) Let $(a_n)_{n\geq 0}$ be a sequence of real numbers. The power series*

$$f(X) = a_0 + a_1 X + \cdots + a_n X^n + \cdots$$

is called the generating function of the sequence, defined for those values of X such that the series converges.

The generating function of a recursive sequence can sometimes be used to find the general term of the sequence. In this case, we say that we solved the recurrence. Recall that the sum of a geometric series is given by

$$1 + X + X^2 + \cdots + X^n + \cdots = \frac{1}{1-X}$$

for $|X| < 1$. By differentiation, we get

$$1 + 2X + 3X^2 + \cdots + (n+1)X^n + \cdots = \frac{1}{(1-X)^2}.$$

Differentiating again, we get

$$\frac{1}{(1-X)^3} = 1 + \frac{3\cdot 2}{2}X + \frac{4\cdot 3}{2}X^2 + \cdots \frac{(n+2)(n+1)}{2}X^n + \cdots .$$

Example 9.33. *Consider the sequence $(a_n)_{n\geq 0}$ with $a_n = 2a_{n-1}, a_0 = 1$. The generating function is*

$$f(X) = a_0 + a_1 X + a_2 X^2 + \cdots + a_n X^n + \cdots = a_0 + 2a_0 X + 2a_1 X^2 + \cdots + 2a_{n-1}X^n + \cdots =$$

$$= a_0 + 2X(a_0 + a_1 X + \cdots + a_{n-1}X^{n-1} + \cdots) = 1 + 2Xf(X),$$

hence

$$f(X) = \frac{1}{1-2X} = 1 + 2X + 2^2 X^2 + \cdots + 2^n X^n + \cdots$$

for $|X| < 1/2$. We get $a_n = 2^n$ for $n \geq 0$.

Example 9.34. *Let $(r_n)_{n\geq 0}$ be the recursive sequence with $r_{n+1} = r_n + n + 1, r_0 = 1$. Then*

$$f(X) = r_0 + r_1 X + r_2 X^2 + \cdots + r_{n+1}X^{n+1} + \cdots =$$

$$= 1 + (r_0 + 1)X + (r_1 + 2)X^2 + \cdots + (r_n + n + 1)X^{n+1} + \cdots =$$

$$= (1 + X + 2X^2 + \cdots + (n+1)X^{n+1} + \cdots) + X(r_0 + r_1 X + \cdots + r_n X^n + \cdots)$$

$$= 1 + \frac{X}{(1-X)^2} + Xf(X)$$

for $|X| < 1$. We get

$$f(X) = \frac{1}{1-X} + \frac{X}{(1-X)^3}.$$

Since

$$\frac{1}{(1-X)^3} = 1 + \frac{3 \cdot 2}{2}X + \frac{4 \cdot 3}{2}X^2 + \cdots \frac{(n+2)(n+1)}{2}X^n + \cdots,$$

we obtain

$$f(X) = 1 + X + X^2 + \cdots X^n + \cdots + X + \frac{3 \cdot 2}{2}X^2 + \frac{4 \cdot 3}{2}X^3 + \cdots \frac{(n+1)n}{2}X^n + \cdots$$

and

$$r_n = 1 + \frac{n(n+1)}{2}.$$

9.5 Exercises

Exercise 9.35. *Suppose* $|A \cup B| = 7, |A| = 5$ *and* $|B| = 4$. *Find* $|A \cap B|$.

Exercise 9.36. *Three sets have* 100 *elements each. Any two of them have exactly* 50 *elements in common. Exactly* 25 *elements are in all three. How many elements are in the union?*

Exercise 9.37. *Assume that* 73% *of British people like cheese,* 76% *like apples, and* 10% *like neither. What percentage of British people like both cheese and apples?*

Exercise 9.38. *Suppose* 2000 *people are at a gathering. Some people have the same birthday. Find the minimum number of such people.*

Exercise 9.39. *a) Find the number of integers between* 1 *and* 5000 *which are not divisible by* 3 *or by* 4.

b) Find the number of integers between 1 *and* 5000 *which are divisible by one or more of the numbers* 4, 5, *and* 6.

Exercise 9.40. *In a certain state, a license plate consists of three letters* $A, B, C, ..., Z$ *followed by three digits from* 0 *to* 9. *How many license plates can be issued?*

Exercise 9.41. *How many integers between* 1 *and* 1000 *have distinct digits?*

Exercise 9.42. *Suppose* 51 *points are chosen inside a* 1×1 *square. Prove that there are at least three points that can be covered by a disc of radius* 1/7.

Exercise 9.43. *Choose* $n + 1$ *integers from the set* $\{1, 2, ..., 2n\}, n \geq 1$. *Prove that two of them are relatively prime.*

Exercise 9.44. *Consider a society of* n *people.*

a) Prove that the number of people with an odd number of friends in this society is even.

b) If everybody has exactly 3 *friends, prove that* n *is even.*

Exercise 9.45. *A rook is placed on the lower left corner of a usual 8×8 chessboard. Is it possible to move the rook such that it visits every square of the chessboard once and only once and such that it ends at the upper right corner?*

Exercise 9.46. *In how many ways can the letters of the word* land *be rearranged? Same question for* mara *and* llama.

Exercise 9.47. *The digits $1, 2, 3, 4, 5, 6$ are written down in some order to form a six-digit number.*
 a) How many such numbers are there?
 b) How many such numbers are even?
 c) How many are divisible by 4? By 8? By 11?

Exercise 9.48. *In how many ways can eight identical rooks be placed on a 8×8 chessboard so that no rook attacks another rook?*

Exercise 9.49. *Prove that $P(n, k)$ is the number of injective functions from $\{1, 2, ..., k\}$ to $\{1, 2, ..., n\}$.*

Exercise 9.50. *Expand the binomials $(2x + y)^3, (2x + 3y)^4, (2x - 3y)^5$.*

Exercise 9.51. *Find a such that*

$$1 + \frac{1}{2}\binom{n}{1} + \frac{1}{3}\binom{n}{2} + \cdots + \frac{1}{n+1}\binom{n}{n} = \frac{a}{n+1}.$$

Hint: Integrate the identity $(1 + x)^n = \sum_{k=0}^{n}\binom{n}{k}x^k$ and then take $x = 1$.

Exercise 9.52. *Twelve balls are placed in a jar. Four are white, three are red, and five are blue. Five balls are taken from the jar. How many selections are possible that contain exactly three white balls?*

Exercise 9.53. *Prove that*

$$\sum_{r=0}^{k}\binom{m}{k-r}\binom{n}{r} = \binom{m+n}{k}.$$

Exercise 9.54. *Try to find a combinatorial proof of the equality*

$$1 + \binom{n}{1}2 + \binom{n}{2}4 + \cdots + \binom{n}{n-1}2^{n-1} + \binom{n}{n}2^n = 3^n.$$

Exercise 9.55. *Solve the recurrence relations:*
 a) $a_n = a_{n-1} + 2a_{n-2}, a_0 = 3, a_1 = 2$.
 b) $a_n = 2a_{n-2}, a_0 = 1, a_1 = 2$.
 c) $2a_n = 3a_{n-1} - a_{n-2}, a_0 = 1, a_1 = 2$.
 d) $a_n = 2a_{n-1} + n/2, a_0 = 1$.

Exercise 9.56. *Use generating functions to solve the recurrence $a_n = 4a_{n-1}, a_0 = 2$.*

Exercise 9.57. *Consider $n \geq 3$ straight lines in the plane such that no two are parallel and exactly three are concurrent. In how many regions is the plane divided?*

Exercise 9.58. ** Let b_n be the number of strings of 0 and 1 of length n having no two consecutive 0's. Find a recurrence relation for b_n and solve it.*

Exercise 9.59. ** The Catalan numbers c_n are defined as the number of triangulations of a polygon with $(n + 2)$ vertices, obtained using diagonals. It is easy to see that $c_1 = 1, c_2 = 2, c_3 = 5$. The recurrence formula for the Catalan numbers is $c_{n+1} = \sum_{i=0}^{n} c_i c_{n-i}$ where $c_0 = 1$.*

a) Compute the generating function $f(X) = \sum_{n=0}^{\infty} c_n X^n$.

b) Derive the formula $c_n = \dfrac{1}{n+1} \dbinom{2n}{n}$ for $n \geq 1$.

10

The construction of integers and rationals

In this chapter we construct the set of integers $\mathbb{Z}$ and the set of rationals $\mathbb{Q}$ using appropriate equivalence relations on pairs of natural numbers and on pairs of integers, respectively. Both equivalence relations are denoted $\sim$ and the equivalence classes will be denoted by $[a, b]$. The fact that a, b are natural numbers or integers should be determined from the context. We define operations on $\mathbb{Z}$ and $\mathbb{Q}$, the usual order relation, and prove several properties. In particular, we show how to identify $\mathbb{N}$ with a subset of $\mathbb{Z}$ and $\mathbb{Z}$ with a subset of $\mathbb{Q}$. We conclude with the decimal representation of the rationals.

10.1   Definition of integers and operations

The invention or discovery of the set of integers $\mathbb{Z} = \{..., -2, -1, 0, 1, 2, 3, ...\}$ was based on pragmatic concerns. Specifically, there were equations, such as $x + 6 = 2$, which mathematicians could not solve using the natural numbers, but which cried out for solution. As a result, a new set of numbers was developed. This set, referred to as the set of integers, was initially considered by many as, at best, a necessary evil. Negative integers were relegated to second-class status and used only when absolutely necessary. Ultimately, however, their usefulness could not be denied and they gradually gained acceptance over the course of the years.

In these more enlightened times, there are lots of ways to model the idea of a negative integer, so that most people can tie the concept down to something more or less concrete. For example, a football buff's attention can be directed towards the idea of a fullback gaining or losing yardage; the accountant type can think in terms of owing money or being owed; the game player can consider the difference between having points or being in the hole (negative numbers often make an appearance on *Jeopardy!*); and the geometrically inclined individual can perceive the difference between motion to the right and motion to the left. Also, the weatherman uses positive and negative temperatures.

It is possible to introduce the concept of the integer by announcing that we are inventing a new kind of number, denoted by ^-n, and specifying how such a number will interact with the already known natural numbers. This can be a bit unwieldy, so we prefer a different approach.

Our introduction to the theory of integers is based on the idea of using pairs of natural numbers to represent integers. Intuitively, the first component of a pair will describe how much plus stuff is involved, while the second component will count the number of minuses. Thus, when we write the pair $\langle 5, 2 \rangle$, we will think of the net result of combining 5 pluses with 2 minuses, which of course is 3 pluses. Likewise $\langle 3, 8 \rangle$ will be thought of as 3 pluses and 8 minuses together, which is equivalent to 5 minuses. Notice that these thoughts entail our regarding of the pairs $\langle 7, 9 \rangle$, $\langle 0, 2 \rangle$, and $\langle 37, 39 \rangle$ as representing the same integer.

To make all of this precise, we will introduce an equivalence relation on the set $\mathbb{N} \times \mathbb{N}$ of ordered pairs of natural numbers and define an integer to be one of the equivalence classes determined by this relation. In addition, we will define the addition, multiplication, subtraction, and division operations. We will then extend the usual order relation $\leq$ from $\mathbb{N}$ to $\mathbb{Z}$.

We begin by defining the relation $\sim$ between ordered pairs of natural numbers that will, in effect, determine when the pairs really represent the same integer.

Definition 10.1. *If $\langle a, b \rangle, \langle c, d \rangle \in \mathbb{N} \times \mathbb{N}$, then $\langle a, b \rangle \sim \langle c, d \rangle$ iff $a + d = b + c$.*

Notice that the objects that are related are themselves ordered pairs. The elements of the actual relation are thus pairs whose components are also ordered pairs. An actual element of the relation $\sim$ on $\mathbb{N} \times \mathbb{N}$ will have the form $\langle \langle a, b \rangle, \langle c, d \rangle \rangle$.

As usual, the definition can be used in two distinct ways. First, if we know that $\langle a, b \rangle \sim \langle c, d \rangle$, then we can immediately assert that $a + d = b + c$. If the hypothesis of a theorem is $\langle 6, b \rangle \sim \langle c, 12 \rangle$, then an immediate conclusion is $6 + 12 = b + c$. If you know that $\langle 4a, a \rangle \sim \langle 10, 4 \rangle$, then you also have $4a + 4 = a + 10$, so $a = 2$.

Second, whenever you see the expression $a + d = b + c$, you may immediately write the equivalent statement $\langle a, b \rangle \sim \langle c, d \rangle$. Thus, since $5 + 4 = 2 + 7$, we have $\langle 5, 2 \rangle \sim \langle 7, 4 \rangle$. Also, if $2a + 1 = 4b + 3$, then $\langle 2a, 3 \rangle \sim \langle 4b, 1 \rangle$. These ideas figure prominently in the proofs of the next few theorems.

Theorem 10.2. *The relation $\sim$ is an equivalence relation on $\mathbb{N} \times \mathbb{N}$.*

Proof. We must separately verify each of the three conditions that appear in the definition of an equivalence relation.

1) Suppose $\langle a, b \rangle \in \mathbb{N} \times \mathbb{N}$. Since $a + b = b + a$, it follows that $\langle a, b \rangle \sim \langle a, b \rangle$. Thus $\sim$ is reflexive.

2) Suppose $\langle a, b \rangle \sim \langle c, d \rangle$. Then $a + d = b + c$. Hence, by the commutative property of addition in $\mathbb{N}$, together with the symmetry property of equality, we get $c + b = d + a$. Thus, by definition, $\langle c, d \rangle \sim \langle a, b \rangle$ and $\sim$ is symmetric.

3) Suppose $\langle a, b \rangle \sim \langle c, d \rangle$ and $\langle c, d \rangle \sim \langle e, f \rangle$. Then $a + d = b + c$ and $c + f = d + e$.

Adding these equations produces $a + d + c + f = b + c + d + e$. Clearly c and d cancel, so $a + f = b + e$. We get $\langle a, b \rangle \sim \langle e, f \rangle$, hence $\sim$ is transitive. $\quad\square$

The equivalence class $[\langle a, b \rangle]$ containing the pair $\langle a, b \rangle$ is denoted simply by $[a, b]$. In other words,

$$[a, b] = \{\langle e, f \rangle : \langle e, f \rangle \sim \langle a, b \rangle\}.$$

Definition 10.3. *Each equivalence class $[a, b] \in \mathbb{N} \times \mathbb{N} / \sim$ will be called an integer. The set of all integers will be called $\mathbb{Z}$ (in case you wonder, the letter Z comes from the German word* Zahl*). Thus:*

$$\mathbb{Z} = \{[a, b] : a, b \in \mathbb{N}\}.$$

Note that each integer has infinitely many representations, in the sense that we can choose different representatives; we can write $[13, 15] = [8, 10] = [259, 261] = [a, a + 2]$, among many others.

Theorem 10.4. *If $[a, b] = [a', b']$ and $[c, d] = [c', d']$, then $[a + c, b + d] = [a' + c', b' + d']$ and $[ac + bd, ad + bc] = [a'c' + b'd', a'd' + b'c']$.*

Proof. Since $a + b' = b + a'$ and $c + d' = d + c'$, we get $a + b' + c + d' = b + a' + d + c'$, hence $[a + c, b + d] = [a' + c', b' + d']$. In order to prove $[ac + bd, ad + bc] = [a'c' + b'd', a'd' + b'c']$, we need to show that $ac + bd + a'd' + b'c' = ad + bc + a'c' + b'd'$. Multiplying the equality $a + b' = b + a'$ by c and d, using distributivity, and adding together we get

$$(1) \quad ac + bd + b'c + a'd = ad + bc + a'c + b'd.$$

Similarly, multiplying the equality $c + d' = d + c'$ by a' and b', we get

$$(2) \quad a'd' + b'c' + a'c + b'd = a'c' + b'd' + a'd + b'c.$$

Adding (1) and (2) we get

$$ac + bd + b'c + a'd + a'd' + b'c' + a'c + b'd = ad + bc + a'c + b'd + a'c' + b'd' + a'd + b'c,$$

and by cancellation, $ac + bd + a'd' + b'c' = ad + bc + a'c' + b'd'$. $\quad\square$

Definition 10.5. *We define the addition and multiplication operations on $\mathbb{Z}$, denoted $\oplus$ and $\odot$ (to distinguish from the old operations $+, \cdot$ on $\mathbb{N}$), by*

$$[a, b] \oplus [c, d] = [a + c, b + d],$$

$$[a, b] \odot [c, d] = [ac + bd, ad + bc],$$

which by the previous theorem do not depend on representatives. After we get used to the new operations, we will of course go back to the usual $+$ and $\cdot$ instead of $\oplus, \odot$.

Using the definitions we can easily prove the following.

Theorem 10.6. *We have*

1. $[a, b] = [0, 0] \Leftrightarrow a = b$.
2. $[a, b] \oplus [0, 0] = [0, 0] \oplus [a, b] = [a, b]$.
3. $[0, 0] \odot [a, b] = [0, 0]$.
4. $[1, 0] \odot [a, b] = [a, b]$.

Theorem 10.7. *The operations $\oplus$ and $\odot$ on $\mathbb{Z}$ have the following properties (here a, b, c, d, e, f are natural numbers):*

1. $[a, b] \oplus [c, d] = [c, d] \oplus [a, b]$ *(commutativity)*.
2. $([a, b] \oplus [c, d]) \oplus [e, f] = [a, b] \oplus ([c, d] \oplus [e, f])$ *(associativity)*.
3. $[a, b] \odot [c, d] = [c, d] \odot [a, b]$ *(commutativity)*.
4. $([a, b] \odot [c, d]) \odot [e, f] = [a, b] \odot ([c, d] \odot [e, f])$ *(associativity)*.
5. $[a, b] \odot ([c, d] \oplus [e, f]) = [a, b] \odot [c, d] \oplus [a, b] \odot [e, f]$ *(distributivity)*.
6. $[a, b] \oplus [c, d] = [a, b] \oplus [e, f] \Rightarrow [c, d] = [e, f]$ *(cancellation)*.
7. *If $[a, b] \neq [0, 0]$ and $[a, b] \odot [c, d] = [a, b] \odot [e, f]$, then $[c, d] = [e, f]$ (cancellation). (Note that the first hypothesis could be written in the form $a \neq b$.)*

Proof. 1. This follows from the commutativity of the addition of natural numbers.

2. This follows from the associativity of the addition of natural numbers.

3. Indeed, $[ac + bd, ad + bc] = [ca + db, cb + da]$.

4. We compute $([a, b] \odot [c, d]) \odot [e, f] = [ac + bd, ad + bc] \odot [e, f] = [ace + bde + adf + bcf, acf + bdf + ade + bcf]$ and $[a, b] \odot ([c, d] \odot [e, f]) = [a, b] \odot [ce + df, cf + de] = [ace + adf + bcf + bde, acf + ade + bce + bdf]$.

5. We have $[a, b] \odot ([c, d] \oplus [e, f]) = [a, b] \odot [c + e, d + f] = [ac + ae + bd + bf, ad + af + bc + be]$, $[a, b] \odot [c, d] \oplus [a, b] \odot [e, f] = [ac + bd, ad + bc] \oplus [ae + bf, af + be] = [ac + bd + ae + bf, ad + bc + af + be]$.

6. From $[a, b] \oplus [c, d] = [a, b] \oplus [e, f]$ we get $[a + c, b + d] = [a + e, b + f]$, hence $a + c + b + f = b + d + a + e$. Using cancellation for natural numbers, we get $c + f = d + e$ and therefore $[c, d] = [e, f]$.

7. Indeed, $[a, b] \neq [0, 0] \Leftrightarrow a + 0 \neq b + 0 \Leftrightarrow a \neq b$. Assuming $a \neq b$, there are two cases: either $a < b$ or $b < a$. From $[a, b] \odot [c, d] = [a, b] \odot [e, f]$ we get $[ac + bd, ad + bc] = [ae + bf, af + be]$, hence $ac + bd + af + be = ad + bc + ae + bf$ and $a(c + f) + b(d + e) = a(d + e) + b(c + f)$. Assuming $a < b$, $b - a \in \mathbb{P}$ and the last equality could be put in the form $(b - a)(d + e) = (b - a)(c + f)$. Using the cancellation for multiplication of natural numbers, we get $d + e = c + f$ and therefore $[c, d] = [e, f]$. The case $b < a$ is similar. $\square$

10.2 Order relation on integers

Definition 10.8. *We define the relations $\prec$ and $\succ$ on $\mathbb{Z}$ by the following:*

    1. $[a, b] \prec [c, d] \Leftrightarrow a + d < b + c$.

    2. $[a, b] \succ [c, d] \Leftrightarrow [c, d] \prec [a, b]$. *We use this notation to distinguish from the old $<$ and $>$ for natural numbers. Again, eventually we will also use $<$ and $>$ for integers.*

Theorem 10.9. *(Law of trichotomy) For all natural numbers a, b, c, d, exactly one of the following is true in $\mathbb{Z}$: $[a, b] \prec [c, d]$ or $[a, b] = [c, d]$ or $[c, d] \prec [a, b]$.*

Proof. This follows from the law of trichotomy in $\mathbb{N}$. □

Theorem 10.10. *(Transitivity) $[a, b] \prec [c, d] \wedge [c, d] \prec [e, f] \Rightarrow [a, b] \prec [e, f]$.*

Proof. We have $a + d < b + c$ and $c + f < d + e$. Adding together, $a + d + c + f < b + c + d + e$. Canceling $c + d$, we get $a + f < b + e$, hence $[a, b] \prec [e, f]$. □

From now on, when we wish to refer to an integer without mentioning an equivalence class, we will use lowercase letters near the end of the alphabet. Sometimes we will be able to prove theorems without having to bring equivalence classes into the picture at all, but when we can't, we are always able to fall back on the definition of an integer.

Theorem 10.11. *For all integers x, y, z we have:*

    1. $(x \succ y) \wedge (y \succ z) \Rightarrow x \succ z$.

    2. $x \prec y \Leftrightarrow x \oplus z \prec y \oplus z$.

    3. $(x \prec y) \wedge ([0, 0] \prec z) \Rightarrow x \odot z \prec y \odot z$.

    4. $(x \prec y) \wedge ([0, 0] \succ z) \Rightarrow x \odot z \succ y \odot z$.

Proof. 1. This follows from transitivity of $<$ on $\mathbb{N}$.

    2. We need to work with representatives: let $x = [a, b], y = [c, d], z = [e, f]$. Then $x \oplus z = [a + e, b + f], y \oplus z = [c + e, d + f]$. We have $a + d < b + c$ and adding $e + f$ both sides, $a + d + e + f < b + c + e + f$, hence $x \oplus z \prec y \oplus z$.

    3. Let $x = [a, b], y = [c, d], z = [e, f]$ with $e > f$. We have $[e, f] = [e - f, 0]$. Without loss of generality we may assume $z = [e, 0]$ with $e > 0$. In this case $x \odot z = [ae, be]$ and $y \odot z = [ce, de]$. From $a + d < b + c$, multiplying both sides by e we get $ae + de < be + ce$, hence $x \odot z \prec y \odot z$.

    4. Let $x = [a, b], y = [c, d], z = [e, f]$. Since $z \prec [0, 0]$, we may assume $z = [0, f]$ with $f > 0$. In this case $x \odot z = [bf, af]$ and $y \odot z = [df, cf]$. From $a + d < b + c$, we get $af + df < bf + cf$, hence $y \odot z \prec x \odot z$. □

Definition 10.12. *An integer x is called* positive *if and only if $x \succ [0, 0]$. An integer x is called* negative *if and only if $x \prec [0, 0]$.*

Corollary 10.13. *If x is any integer, then exactly one of the following is true: x is positive, x is negative, or $x = [0, 0]$.*

Proof. Indeed, if $x = [a, b]$, then either $a < b$, $b < a$, or $a = b$. □

Theorem 10.14. *If x and y are positive integers, then $x \oplus y$ and $x \odot y$ are positive integers. For any $x \in \mathbb{Z}$ there is a unique $y \in \mathbb{Z}$ such that $x \oplus y = [0, 0]$. In fact $[a, b] \oplus [b, a] = [0, 0]$ for all $a, b \in \mathbb{N}$.*

Proof. If $x = [a, 0]$ and $y = [c, 0]$ with $a > 0$ and $c > 0$, then $x \oplus y = [a + c, 0]$ and $x \odot y = [ac, 0]$, hence $x \oplus y$ and $x \odot y$ are positive.

For $x = [a, b]$, we can take $y = [b, a]$ and $x \oplus y = [a + b, a + b] = [0, 0]$. If y' is another integer with $x \oplus y' = [0, 0] = x \oplus y$, by cancellation we get $y = y'$. $\qquad\square$

Definition 10.15. *The class $[b, a]$ is called the additive inverse or* opposite *of $[a, b]$ and is denoted by $\ominus[a, b]$. We define the subtraction operation on $\mathbb{Z}$ by $x \ominus y = x \oplus (\ominus y)$.*

We have:
1. $\forall x, y \in \mathbb{Z} \ \exists! z \in \mathbb{Z}$ such that $x = y \oplus z$, namely $z = x \ominus y$.
2. For all $x \in \mathbb{Z}$, $\ominus(\ominus x) = x$.

Theorem 10.16. *For $a, b \in \mathbb{N}$ we have:*
1. $a > b \Rightarrow \exists! n \in \mathbb{P}$ such that $[a, b] = [n, 0]$.
2. $a < b \Rightarrow \exists! m \in \mathbb{P}$ such that $[a, b] = [0, m]$.

Proof. We can take $n = a - b$ in the first case, and $m = b - a$ in the second. $\qquad\square$

Definition 10.17. *If A is a subset of $\mathbb{Z}$, we say that $\mathbb{N}$ is* isomorphic *with A if and only if there exists a one-to-one function f with domain $\mathbb{N}$ and range A such that for all $a, b \in \mathbb{N}$ we have*

$$f(a + b) = f(a) \oplus f(b), \quad f(a \cdot b) = f(a) \odot f(b), \quad and \ a < b \Rightarrow f(a) \prec f(b).$$

Such a function f is called an isomorphism *between $\mathbb{N}$ and A.*

Theorem 10.18. *If $A = \{[a, 0] : a \in \mathbb{N}\}$, then $\mathbb{N}$ is isomorphic with A.*

Proof. Define $f : \mathbb{N} \to A, f(n) = [n, 0]$. Then f is a bijection and preserves addition, multiplication, and strict order. $\qquad\square$

Because there is an isomorphism between $\mathbb{N}$ and the set of nonnegative integers, we will henceforth consider the class $[a, 0]$ to be the same as the natural number a unless there is something to be gained by making a distinction between them. In addition, if x, y are integers, from now on we will ordinarily write: $x + y$ for $x \oplus y$, $x \cdot y$ or xy for $x \odot y$, $x < y$ for $x \prec y$, $-x$ for $\ominus x$, etc.

Definition 10.19. *The set of all positive integers will be denoted by $\mathbb{Z}^+$ or $\mathbb{P}$. The set of all negative integers will be denoted by $\mathbb{Z}^-$.*

Using the properties of $\mathbb{P}$ and $\mathbb{N}$, we obtain the following:

Corollary 10.20. *We have*
1. *If $A \subseteq \mathbb{Z}^+$, $1 \in A$, and $x \in A \Rightarrow x + 1 \in A$, then $G = \mathbb{Z}^+$.*
2. *If $A \subseteq \mathbb{Z}^+$ and $A \neq \emptyset$, then A contains a unique smallest member.*
In part 1, the conclusion remains valid if $\mathbb{Z}^+$ is replaced by the set of all nonnegative integers $\mathbb{Z}^+ \cup \{0\} = \mathbb{N}$ and the hypothesis $1 \in A$ is changed to $0 \in A$.

10.3   Definition of rationals, operations and order

People have worked with fractions for a long time. If you want to share a loaf of bread with three other people, each will get one quarter. Rational numbers were invented since an equation like $3x = 2$ has no integer solution: you need the fraction $2/3$. The modern definition of rational numbers m/n uses an equivalence relation on the set of pairs $\langle m, n \rangle$ where m, n are integers and $n \neq 0$. This construction is generalized in abstract algebra to obtain the so-called field of fractions of an integral domain.

Definition 10.21. *Let* $\mathbb{Z}^* = \mathbb{Z} \setminus \{0\}$. *Define a relation on* $\mathbb{Z} \times \mathbb{Z}^*$ *by*

$$\langle a, b \rangle \sim \langle c, d \rangle \text{ if and only if } ad = bc.$$

Theorem 10.22. *The relation* $\sim$ *is an equivalence relation on* $\mathbb{Z} \times \mathbb{Z}^*$.

Proof. The relation is reflexive $\langle a, b \rangle \sim \langle a, b \rangle$, since $ab = ba$. It is symmetric since $\langle a, b \rangle \sim \langle c, d \rangle$ implies $\langle c, d \rangle \sim \langle a, b \rangle$. Indeed, $ad = bc$ implies $cb = da$. For transitivity, assume $\langle a, b \rangle \sim \langle c, d \rangle$ and $\langle c, d \rangle \sim \langle e, f \rangle$, hence $ad = bc$ and $cf = de$. Multiplying the first equation by f, we get $adf = bcf = bde$. Canceling $d \neq 0$ we get $af = be$, hence $\langle a, b \rangle \sim \langle e, f \rangle$. □

Definition 10.23. *The set of rational numbers* $\mathbb{Q}$ *is the set of equivalence classes determined by* $\sim$ *on* $\mathbb{Z} \times \mathbb{Z}^*$. *The class containing* $\langle a, b \rangle$ *is denoted by* $[a, b]$ *(later this will be denoted in the more traditional fashion by* $\dfrac{a}{b}$ *or by* a/b*). The integer* a *is called the numerator, and the integer* $b \neq 0$ *is called the denominator.*

Theorem 10.24. *With the above notation, we have*
  1. $[a, b] = [c, d] \Leftrightarrow ad = bc$.
  2. *If* $[a, b] = [c, d]$ *and* $[e, f] = [g, h]$, *then:*
      a. $[af + be, bf] = [ch + dg, dh]$ *and*
      b. $[ae, bf] = [cg, dh]$.

Proof. The first part follows from the definition. For the second, assume $ad = bc$ and $eh = fg$. For part a, we compute $(af + be)dh = adfh + bdeh = bcfh + bdfg = bf(ch + dg)$, hence $[af + be, bf] = [ch + dg, dh]$. For part b we just multiply the two equations to get $adeh = bcfg$, hence $[ae, bf] = [cg, dh]$. □

Corollary 10.25. *We define the addition and multiplication on* $\mathbb{Q}$ *by*

$$[a, b] \oplus [c, d] = [ad + bc, bd]$$

and

$$[a, b] \odot [c, d] = [ac, bd].$$

This definition does not depend on representatives.

Theorem 10.26. *We have*
1. $[a, b] = [c, b] \Leftrightarrow a = c$.
2. *If $a \neq 0$, then $[a, b] = [a, d] \Leftrightarrow b = d$.*
3. *We have $[0, 1] \oplus [a, b] = [a, b]$ and $[0, 1] \odot [a, b] = [0, 1]$*
4. *The operations $\oplus$ and $\odot$ are commutative and associative.*
5. *The multiplication $\odot$ is distributive with respect to the addition $\oplus$.*
6. *For $x, y, z \in \mathbb{Q}$ we have $x \oplus y = x \oplus z \Rightarrow y = z$ (cancellation).*
7. *If $x, y, z \in \mathbb{Q}$ and $x \neq [0, 1]$, then $x \odot y = x \odot z \Rightarrow y = z$ (cancellation).*
8. *For each $x \in \mathbb{Q}$ there is a unique $y \in \mathbb{Q}$ such that $x \oplus y = [0, 1]$.*

Proof. 1. Assuming $ab = bc$, we cancel b and get $a = c$. The other implication is easy.

2. If $[a, b] = [a, d]$, we have $ad = ab$ and we can cancel a since $a \neq 0$. The other implication is obvious.

3. By definition $[0, 1] \oplus [a, b] = [0 \cdot b + 1 \cdot a, 1 \cdot b] = [a, b]$. Also, $[0, 1] \odot [a, b] = [0 \cdot a, 1 \cdot b] = [0, b] = [0, 1]$.

4. Straightforward computations:

$$[a, b] \oplus [c, d] = [ad + bc, bd] = [c, d] \oplus [a, b],$$

$$[a, b] \oplus ([c, d] \oplus [e, f]) = [a, b] \oplus [cf + de, df] = [adf + bcf + bde, bdf],$$

$$([a, b] \oplus [c, d]) \oplus [e, f] = [ad + bc, bd] \oplus [e, f] = [adf + bcf + bde, bdf],$$

$$[a, b] \odot [c, d] = [ac, bd] = [c, d] \odot [a, b],$$

$$[a, b] \odot ([c, d] \odot [e, f]) = [a, b] \odot [ce, df] = [ace, bdf],$$

$$([a, b] \odot [c, d]) \odot [e, f] = [ac, bd] \odot [e, f] = [ace, bdf].$$

5. We have

$$[a, b] \odot ([c, d] \oplus [e, f]) = [a, b] \odot [cf + de, df] = [acf + ade, bdf]$$

and

$$([a, b] \odot [c, d]) \oplus ([a, b] \odot [e, f]) = [ac, bd] \oplus [ae, bf] =$$

$$[abcf + abde, b^2 df] = [acf + ade, bdf].$$

6. Let $x = [a, b]$, $y = [c, d]$ and $z = [e, f]$. From $x \oplus y = x \oplus z$ we get $[ad + bc, bd] = [af + be, bf]$, so $[adf + bcf, bdf] = [adf + bde, bdf]$. From 1 we get $adf + bcf = adf + bde$, hence by cancelation of integers, $cf = de$ and therefore $y = z$.

7. Let $x = [a, b]$ with $a \neq 0$, $y = [c, d]$ and $z = [e, f]$. From $x \odot y = x \odot z$ we get $[ac, bd] = [ae, bf]$, so $[acf, bdf] = [ade, bdf]$. From 1 it follows that $acf = ade$, hence $cf = de$ since $a \neq 0$ and therefore $y = z$.

8. For $x = [a, b]$, we can take $y = [-a, b]$, so such a y exists. If $a \oplus y = x \oplus y' = [0, 1]$, by cancellation we get $y = y'$, so y is unique. $\quad\square$

Definition 10.27. *The rational number y such that $x \oplus y = [0,1]$ is called the additive inverse or opposite of x and is denoted by $\ominus x$. We define the operation of subtraction on $\mathbb{Q}$ by $z \ominus x = z \oplus (\ominus x)$.*

Theorem 10.28. *We have*
  1. *$\ominus[a,b] = [-a,b] = [a,-b]$.*
  2. *$[a,b] \ominus [c,d] = [ad - bc, bd]$.*
  3. *$[1,1] \odot [a,b] = [a,b]$.*
  4. *For each $x \in \mathbb{Q}$ with $x \neq [0,1]$ there is a unique $y \in \mathbb{Q}$ such that $x \odot y = [1,1]$.*

Proof. Parts 1–3 follow from straightforward computation. For 4, let $x = [a,b]$ with $a \neq 0$. We can take $y = [b,a]$. Uniqueness follows from cancellation. □

Definition 10.29. *The rational number y such that $x \odot y = [1,1]$ is called the reciprocal or multiplicative inverse of x. It is denoted by x^{-1} or $1/x$. We define the operation of division on $\mathbb{Q}$ by $x \div y = x \odot y^{-1}$ for $y \neq 0$. We also write x/y for $x \div y$.*

Theorem 10.30. *We have*
  1. *For $[a,b] \in \mathbb{Q}$ with $a \neq 0$ we have $[a,b]^{-1} = [b,a]$.*
  2. *For $c \neq 0$ we have $[a,b]/[c,d] = [ad, bc]$.*

Proof. 1. This follows from the computation $[a,b] \odot [b,a] = [ab, ab] = [1,1]$.
  2. Indeed, $[a,b]/[c,d] = [a,b] \odot [c,d]^{-1} = [a,b] \odot [d,c] = [ad, bc]$. □

Definition 10.31. *We say that $[a,b] \in \mathbb{Q}$ is positive if and only if $ab > 0$ and that $[a,b]$ is negative if and only if $ab < 0$. We write $x \succ y$ if and only if $x \ominus y$ is positive, and $x \prec y$ iff $y \succ x$. We write $x \preceq y$ if $x \prec y$ or $x = y$.*

Theorem 10.32. *We have*
  1. *$[a,b]$ is positive iff $[a,b] \succ [0,1]$.*
  2. *$[a,b]$ is negative iff $[-a,b]$ is positive.*
  3. *If x and y are positive, then $x \odot y$ and $x \oplus y$ are positive.*
  4. *(Trichotomy) Exactly one of the following is true: $x \prec y$, $y \prec x$, or $x = y$.*
  5. *(Transitivity) $x \prec y$ and $y \prec z$ implies $x \prec z$.*
  6. *$x \prec y \Rightarrow x \oplus z \prec y \oplus z$.*
  7. *$(x \prec y) \wedge (z \prec w) \Rightarrow x \oplus z \prec y \oplus w$.*
  8. *$([0,1] \prec x) \wedge (y \prec z) \Rightarrow x \odot y \prec x \odot z$.*

Proof. 1. Indeed, $[a,b] \ominus [0,1] = [a,b]$ is positive iff $ab > 0$.
  2. We have $ab < 0 \Leftrightarrow (-a)b > 0$.
  3. Let $x = [a,b]$ and $y = [c,d]$ with $ab > 0, cd > 0$. Then $x \odot y = [ac, bd]$ and $acbd > 0$. Also, $x \oplus y = [ad + bc, bd]$ and $(ad + bc)bd = abd^2 + b^2cd > 0$.
  4. Let $x \ominus y = [a,b] \in \mathbb{Q}$. Then we have either $ab = 0$ or $ab > 0$ or $ab < 0$. In the first case, $x = y$, in the second, $x \prec y$, and in the third, $y \prec x$.

5. If $y \ominus x = [a, b]$ and $z \ominus y = [c, d]$ are positive, then $z \ominus x = [a, b] \oplus [c, d]$ is also positive.

6. Assume $x \prec y$. We have $(y \oplus z) \ominus (x \oplus z) = y \ominus x$, which is positive, hence $x \oplus z \prec y \oplus z$.

7. Assume $x \prec y$ and $z \prec w$. Then $(y \oplus w) \ominus (x \oplus z) = (y \ominus x) \oplus (w \ominus z)$ is positive since $y \ominus x$ and $w \ominus z$ are positive.

8. Assume $[0, 1] \prec x$ and $y \prec z$. Then $(x \odot z) \ominus (x \odot y) = x \odot (z \ominus y)$ is positive, as the product of positives. □

Definition 10.33. *We say that a pair $\langle a, b \rangle \in \mathbb{Z} \times \mathbb{Z}^*$ is in lowest terms iff $b > 0$ and $gcd(a, b) = 1$.*

Theorem 10.34. *Every nonzero rational number can be represented uniquely by a pair in lowest terms.*

Proof. Let $x = [a, b]$ with $a \neq 0$ and $b > 0$. Let $d = \gcd(a, b)$ and let $a' = a/d, b' = b/d$. Then $[a, b] = [da', db'] = [a', b']$, where $b' > 0$ and $\gcd(a', b') = 1$. The numbers a', b' are unique. Indeed, let $[a'', b'']$ be another representation of $[a, b]$ in lowest terms. Then $[a', b'] = [a'', b'']$ implies $a'b'' = a''b'$, hence $b' \mid a'b''$. Since $\gcd(a', b') = 1$, it follows from Lemma 7.12 that $b' \mid b''$. A similar argument shows that $b'' \mid b'$, hence $b' = b''$ because they are positive. We also get $a' = a''$, hence we have uniqueness. □

Theorem 10.35. *The subset $W = \{[a, 1] : a \in \mathbb{Z}\} \subset \mathbb{Q}$ can be identified with $\mathbb{Z}$. We say that W is isomorphic with $\mathbb{Z}$.*

Proof. Define $f : \mathbb{Z} \to W, f(a) = [a, 1]$. Then f is one-to-one and onto, $f(a + b) = f(a) \oplus f(b), f(ab) = f(a) \odot f(b)$ and $a \leq b \Rightarrow f(a) \preceq f(b)$. □

From now on, we will write $+, -, \cdot, \leq$ instead of $\oplus, \ominus, \odot, \preceq$. Also, a rational number $[a, b]$ will be denoted by $\dfrac{a}{b}$ or a/b. An integer k is identified with the rational number $[k, 1] = k/1$. The notation a/b for division of rational numbers is consistent with the case when a, b are integers.

Corollary 10.36. *In particular we have the (strict) inclusions $\mathbb{N} \subset \mathbb{Z} \subset \mathbb{Q}$.*

Theorem 10.37. *Between any two distinct rational numbers there is another rational number.*

Proof. Suppose $r, s \in \mathbb{Q}$ with $r < s$ and consider $\dfrac{r + s}{2} \in \mathbb{Q}$. Then $r < \dfrac{r + s}{2} < s$ since $r + r < r + s < s + s$. □

Theorem 10.38. *(Archimedean property) If r, s are positive rational numbers, then there is a positive integer n such that $nr > s$.*

Proof. Let $r = a/b, s = c/d$, where a, b, c, d are positive integers. For n an arbitrary positive integer, the inequality $nr > s$ is equivalent with $\dfrac{na}{b} > \dfrac{c}{d}$ or $nad > bc$. We may pick $n = 2bc$ and the inequality is satisfied since $ad \geq 1$. □

Definition 10.39. *The absolute value of a rational number is defined as*

$$|x| = \begin{cases} x & if \quad x \geq 0 \\ -x & if \quad x < 0. \end{cases}$$

We summarize the properties of the rational numbers by saying that $\mathbb{Q}$ together with the operations $+, \cdot$ and the natural order relation $\leq$ is an ordered field with the Archimedean property, containing the set of integers $\mathbb{Z}$. Since the equation $x^2 = 2$ has no rational solution, the field $\mathbb{Q}$ is not complete. It is riddled with "holes" that will be filled by constructing the field of real numbers in the next chapter.

10.4 Decimal representation of rational numbers

A decimal fraction is a rational number of the form $\dfrac{a}{10^n}$ where $a \in \mathbb{Z}$ and $n \geq 1$. For example, $\dfrac{-17}{10^3}$ is a decimal fraction equal to the decimal number -0.017. You are already familiar with decimal numbers, say from a pocket calculator. When the decimals continue forever, a decimal number $0.a_1 a_2 ... a_n ...$ where $a_n \in \{0, 1, ..., 9\}$ is the sum of the series

$$\frac{a_1}{10} + \frac{a_2}{10^2} + \cdots + \frac{a_n}{10^n} + \cdots$$

Each rational number $\dfrac{a}{b}$ can be represented as a decimal number with finitely many nonzero decimals or with repeating decimals. For example

$$\frac{1}{2} = \frac{5}{10} = 0.5, \quad \frac{5}{6} = 0.8333... = \frac{8}{10} + \frac{3}{10^2} + \frac{3}{10^3} + \cdots = 0.8\bar{3},$$

$$-\frac{2}{7} = -0.285714285714... = \frac{-2}{10} + \frac{-8}{10^2} + \frac{-5}{10^3} + \frac{-7}{10^4} + \frac{-1}{10^5} + \frac{-4}{10^6} + \cdots = -0.\overline{285714}.$$

We overline the group of repeating decimals. Usually, the trailing zeros are omitted. The decimal representation is not unique, for example,

$$\frac{1}{5} = 0.2 = 0.19999... = 0.1\bar{9}.$$

Indeed,

$$\frac{1}{10} + \frac{9}{10^2} + \frac{9}{10^3} + \cdots = \frac{1}{10} + \frac{9}{100} \sum_{k=0}^{\infty} \frac{1}{10^k} = \frac{1}{10} + \frac{9}{100} \cdot \frac{1}{1 - \frac{1}{10}} = \frac{2}{10} = \frac{1}{5}.$$

Here we used the sum of the geometric series $\displaystyle\sum_{k=0}^{\infty} r^k = \frac{1}{1 - r}$ for $|r| < 1$.

If we avoid repeated nines, then the decimal representation is unique. In fact, the following is true.

Theorem 10.40. *Any rational number in lowest terms such that the only prime factors of the denominator are 2 and/or 5 can be expressed as a decimal fraction and has a finite decimal representation.*

Any rational number in lowest terms such that its denominator has prime factors other than 2 and 5 has a unique infinite periodic decimal representation and the period starts right after the decimal point.

If the denominator has prime factors 2 or 5 and other primes, then it has a unique infinite periodic decimal representation where the period starts later.

Proof. Of course, the integers have no nonzero decimals (we exclude the repeated nines), so we assume that we deal with rational numbers which are not integers. Suppose $\frac{a}{b}$ is in lowest terms and the only prime factors of b are 2 and 5, say $b = 2^m 5^n$. If $m = n$, then we are done, since $\frac{a}{b} = \frac{a}{10^n}$. If $m < n$, then by multiplying both the numerator and the denominator of $\frac{a}{b}$ by 2^{n-m}, we get $\frac{a}{b} = \frac{2^{n-m}a}{10^n}$. Similarly, if $m > n$, then $\frac{a}{b} = \frac{5^{m-n}}{10^m}$.

Suppose now that $\frac{a}{b}$ is in lowest terms and that b is not divisible by 2 or by 5. We may assume $0 < a < b$. The decimals are obtained by the long division algorithm,

$$a = b \cdot 0 + a, \quad 10a = b \cdot q_1 + r_1, \quad 10r_1 = b \cdot q_2 + r_2, \ldots$$

Since there are at most $b - 1$ possible remainders, the first remainder a will reappear after at most $b - 1$ steps, forcing the decimals to repeat from that step on. The length of the period is at most $b - 1$.

Finally, if $\frac{a}{b}$ is in lowest terms and b has prime factors 2 and/or 5 together with other primes, then $b = 2^m 5^n b'$ where b' is not divisible by 2 or by 5. If $m \leq n$ and $n \geq 1$, then $10^n \frac{a}{b} = \frac{a'}{b'}$ with a', b' relatively prime, and as above $\frac{a'}{b'}$ is a periodic decimal number. To get back $\frac{a}{b}$, we divide by 10^n. This amounts to a shift of the decimal point, obtaining a decimal number where the period starts n digits after the decimal point. The case $m > n$ is similar. □

We can use any base $b \geq 2$ to express rational numbers as a finite or infinite sum of fractions of the form $\frac{a_n}{b^n}$ with $a_n \in \{0, 1, \ldots, b - 1\}$, where the prime factors of b play a similar role as 2 and 5 in the case $b = 10$.

Example 10.41. *We can express $\frac{2}{7}$ as an infinite sum of fractions of the form $\frac{a_n}{9^n}$, using the base $b = 9 = 3 \cdot 3$. Indeed,*

$$\frac{2}{7} = \frac{2}{9} + \frac{5}{9^2} + \frac{1}{9^3} + \frac{2}{9^4} + \frac{5}{9^5} + \frac{1}{9^6} + \cdots = 0.\overline{251} \text{ in base 9.}$$

Here the period $\overline{251}$ starts right away since 7 is not divisible by 3.

10.5 Exercises

Exercise 10.42. *Let $[a, b]$ denote the equivalence class used to define $\mathbb{Z}$.*
  a. Show that $\langle 6, 12 \rangle \in [8, 14]$.
  b. Prove: if $\langle c, c \rangle \in [a, b]$, then $a = b$.
  c. Prove: if $\langle c, c + 5 \rangle \in [4, b]$, then $b = 9$.

Exercise 10.43. *a. Show that the relation $\prec$ is well defined on $\mathbb{Z}$, in other words, prove that if $\langle a, b \rangle \sim \langle a', b' \rangle$, $\langle c, d \rangle \sim \langle c', d' \rangle$, and $a + d < b + c$, then $a' + d' < b' + c'$.*
  b. Suppose we defined $[a, b] \prec [c, d]$ to mean $a < c$. Would this be well-defined? Why or why not?

Exercise 10.44. *Prove that for $x, y, z \in \mathbb{Z}$:*
  a. $\ominus(x \oplus y) = (\ominus x) \oplus (\ominus y)$.
  b. $\ominus(x \ominus y) = y \ominus x$.
  c. $x \odot (\ominus y) = (\ominus x) \odot y = \ominus(x \odot y)$.
  d. $(\ominus x) \odot (\ominus y) = x \odot y$.
  e. $[0, 1] \odot x = \ominus x$.
  f. $x \odot y = [0, 0] \Rightarrow (x = [0, 0]) \vee (y = [0, 0])$.
  g. $x \odot (y \ominus z) = (x \odot y) \ominus (x \odot z)$.

Exercise 10.45. *For each of the following functions $f : \mathbb{N} \to \mathbb{Z}$, determine if f defines an isomorphism of $\mathbb{N}$ onto the range $A = f(\mathbb{N}) \subseteq \mathbb{Z}$.*
  a. $f(a) = [a, a]$.
  b. $f(a) = [a, a + 1]$.
  c. $f(a) = [a, 2a]$.
  d. $f(a) = [0, a]$.
  e. $f(a) = [a + 1, 2]$.
  f. $f(a) = [2a, a]$.
  g. $f(a) = [5a, 0]$.

Exercise 10.46. *On the set $\mathbb{Z} \times \mathbb{Z}$ define the relation $\langle a, b \rangle \sim \langle c, d \rangle$ if $a + d = b + c$. Show that $\sim$ is an equivalence relation and that the quotient set $(\mathbb{Z} \times \mathbb{Z})/\sim$ can be identified with $\mathbb{Z}$.*

Exercise 10.47. *Is $f : \mathbb{Q} \to \mathbb{Z}, f([a, b]) = a - b$ a well-defined function? (Hint: Check if the formula depends on representatives).*

Exercise 10.48. *Prove that between two rational numbers there are infinitely many rationals.*

Exercise 10.49. *For $x, y \in \mathbb{Q}$ we have the properties.*
  a. $|x| \geq 0$ and $|x| = \max(x, -x)$.
  b. $|x + y| \leq |x| + |y|$.
  c. $|xy| = |x||y|$.
  d. $\big||x| - |y|\big| \leq |x - y|$.

Exercise 10.50. *On* $X = \mathbb{Q} \times (\mathbb{Q} \setminus \{0\})$ *define the relation* $\langle a, b \rangle \sim \langle c, d \rangle$ *if* $ad = bc$. *Show that* $\sim$ *is an equivalence relation and that the quotient set* $X/\sim$ *can be identified with* $\mathbb{Q}$.

Exercise 10.51. *Write the following fractions as decimal numbers and find their period*

$$\frac{1}{9}, \quad \frac{122}{99}, \quad \frac{34471}{99900}.$$

Exercise 10.52. *Express* $\dfrac{4}{21}$ *as a sum of fractions of the form* $\dfrac{a_n}{9^n}$ *where* $a_n \in \{0, 1, ..., 8\}$. *Find its period.*

11

The construction of real and complex numbers

In this chapter we describe two different methods to construct the set of real numbers $\mathbb{R}$, the Dedekind cuts approach, and the Cauchy sequence approach. Both methods "fill" the gaps in the set of rational numbers. We define operations and the natural order relation on $\mathbb{R}$ and prove several important properties, in particular that $\mathbb{R}$ is a complete ordered field with the Archimedean property. We describe the decimal representation of real numbers. We also define algebraic and transcendental numbers.

We construct the complex numbers $\mathbb{C}$ starting with pairs of real numbers, define operations, and show some geometric properties, including the trigonometric form of a complex number. If the real numbers can be thought of as the points of a line, the complex numbers can be represented as points in the plane. The field of complex numbers is also complete, but it has no order relation compatible with the operations.

11.1 The Dedekind cuts approach

Examples of real numbers were discovered quite early in the mathematical game. In fact, the existence of $\sqrt{2}$ is implied by the Pythagorean theorem, which has been known for well over 2000 years. We already proved that $\sqrt{2}$ is not rational. In the decimal representation, $\sqrt{2} = 1.41421356237...$ you notice no periodicity, no matter how far you may go. The same will happen for the decimal representation of the number π, the length of a circle of radius $1/2$. The fact that π is irrational is more difficult to prove. In fact, an irrational number can be thought of as an infinite decimal number with no periodicity. We will make this more precise later.

Our first approach to the construction of real numbers is based on ideas first propounded by the German mathematician R. Dedekind more than 100 years ago, called the *Dedekind cuts*. The idea of a cut is to split the rational numbers in two halves whenever there is a "hole".

As in the construction of integers and rationals, we begin with a specific system of numbers, in this case $\mathbb{Q}$, and construct a new number system from it. We will freely use any known properties of rationals as we try to prove theorems about reals. We denote by $\mathbb{Q}^+$ the set of positive rational numbers, and by $\mathbb{Q}^-$ the set of negative rational numbers.

There is an intuitive idea underlying the notion of a real number, namely, that each real number is uniquely determined by its position relative to the rational numbers. In other words, if we are given a number and we know exactly which rationals are less than it, then the number is not only known, it can be calculated to an arbitrary degree of precision using rational approximations. Thus, in order to define certain objects which will eventually be called real numbers, we shall begin by considering sets of rationals that, in effect, can be thought of as consisting of all rationals "to the left of" somewhere. Of course, we could choose to talk about all rational numbers "to the right of somewhere", but the situation would be perfectly symmetric. These sets of rationals will be called *cuts*, and their formal definition is below.

Definition 11.1. *Suppose $C \subseteq \mathbb{Q}$. We say that the set C is a* Dedekind cut *if and only if:*
1. *$C \neq \emptyset$,*
2. *$C \neq \mathbb{Q}$,*
3. *$x \in C \Rightarrow \exists y \in C$ with $x < y$,*
4. *$x \in C$ and $y \in \mathbb{Q}$ with $y < x$ implies $y \in C$.*

If C is a cut, then we know four things about C. It contains at least one rational number, because it is nonempty. The fact that C is not equal to the entire set $\mathbb{Q}$ guarantees that there is at least one rational number which is not in C. Condition 3 asserts that C contains no largest member, because no matter what element of C we might choose, there always exists another number in C which is greater. Finally, if x is known to be in C, then every number less than x is also in C.

On the other hand, if we wish to prove that a certain set is a cut, we must establish that all four of the listed criteria are satisfied for the set. This means, for example, that to verify property 1 we must actually describe a rational number in the set or else show that the assumption "the set is empty" leads to a contradiction.

Example 11.2. *Let $C = \{x \in \mathbb{Q} : x < 7\}$. We claim that C is a cut.*

Proof. To back up the claim we provide a proof in four parts.

1. We need to show that C is nonempty, and this can be done simply by taking a wild guess and checking that the guess is correct. Since $3 < 7$, we see by definition that $3 \in C$. Thus $C \neq \emptyset$. There is nothing special about 3. If we had selected -387, $17/16$, or any other definite rational number less than 7 we would have drawn the same conclusion.

2. In a similar way, as soon as we note that, for example, $8 \notin C$, we have proved that $C \neq \mathbb{Q}$.

3. Since this property is phrased as a conditional, we must begin by examining some arbitrary but definite element of C. Suppose $x \in C$. Then $x < 7$.

How can we find something in C which is greater than x? Here, of course, we cannot write down a definite number value since we don't know exactly what x is. But we can do the next best thing. We can write a formula in terms of x which will always generate a number of the desired sort. Recalling that the arithmetic mean of two numbers lies between them is helpful here. In fact, suppose we let $y = \frac{x+7}{2}$. Then certainly $x < y$ and $y < 7$, so $y \in C$.

4. Suppose $x \in C$, $y \in \mathbb{Q}$, and $y < x$. By the definition of C, we have $x < 7$, so by transitivity it is also true that $y < 7$. Hence $y \in C$. $\qquad\square$

The argument given above can obviously be modified to apply to any similar type of sets of the form $\{y \in \mathbb{Q} : y < r\}$.

Corollary 11.3. *For each $r \in \mathbb{Q}$, the set $\{y \in \mathbb{Q} : y < r\}$ is a cut.*

Definition 11.4. *For each $r \in \mathbb{Q}$, we define the cut $\hat{r} = \{y \in \mathbb{Q} : y < r\}$.*

The following example is a different type of cut.

Example 11.5. *The set*

$$D = \{x \in \mathbb{Q}^+ : x^2 < 2\} \cup \{x \in \mathbb{Q} : x \le 0\}$$

is a cut.

Proof. 1. Since $0 \in D$, D is nonempty.

2. For example $2 \notin D$ since $2^2 = 4 > 2$, hence $D \neq \mathbb{Q}$.

3. Let $x \in D$ and let us find $y \in D$ with $x < y$. If $x \le 0$, we can take $y = 1$. Let $x > 0$ with $x^2 < 2$ and consider $y = x + \dfrac{2 - x^2}{x + 2} = \dfrac{2x + 2}{x + 2}$. It is clear that $y \in \mathbb{Q}$ and that $y > x$. An easy computation shows that $y^2 - 2 = \dfrac{2x^2 - 4}{(x+2)^2} < 0$, hence $y \in D$.

4. If $x \in D$ and $r < x$, then there are two cases: $r > 0$ or $r \le 0$. In the first case, $x > 0$ and $r^2 < x^2 < 2$, so $r \in D$. In the second case, clearly $r \in D$. $\qquad\square$

Definition 11.6. *If C and D are cuts, then we write $C \prec D$ iff $C \subset D$ (strict inclusion). Also, we write $C \preceq D$ if $C \prec D$ or $C = D$.*

Theorem 11.7. *We have*

1. Let C be a cut. If $x \notin C$, then for all $y \in C$ we have $y < x$.

2. For all cuts C and D, exactly one of the following is true: $C \prec D, C = D$ or $D \prec C$ (trichotomy).

3. $(C \prec D) \wedge (D \prec E) \rightarrow C \prec E$ (transitivity).

4. $C \prec D \Rightarrow \exists B$ with $C \prec B \prec D$.

Proof. 1. Suppose the conclusion is false. Then there is a $y \in C$ such that $x \le y$. But $x < y$ is false by part 4 of the definition of a cut and $x = y$ is false since $x \notin C$. Either way we have a contradiction.

Parts 2 and 3 follow from the properties of set inclusion.

4. By hypothesis, there is an r in D such that $r \notin C$. Use property 3 of a cut to find an element r' of D such that $r < r'$. We claim that $C \prec \hat{r'} \prec D$, which means that we may take $B = \hat{r'}$. Indeed, since $r < r'$, we clearly have $r \in \hat{r'}$. This, together with the fact that $r \notin C$, tells us that $C \prec \hat{r'}$. Let $r'' \in D$ such that $r' < r''$. Then clearly $r'' \notin \hat{r'}$, so by definition $\hat{r'} \prec D$.

Note that the second part of the previous proof could be made easier by observing that r' itself is not an element of $\hat{r'}$. $\qquad\square$

Definition 11.8. *A cut C is positive if $\exists r \in \mathbb{Q}^+$ with $r \in C$. A cut C is negative if $\exists r \in \mathbb{Q}^-$ with $r \notin C$.*

Theorem 11.9. *If C is a cut, then exactly one of the following is true: C is positive, C is negative, or $C = \hat{0}$.*

Proof. Suppose $C \neq \hat{0}$. If there is $r \in \mathbb{Q}^+$ with $r \in C$, then C is positive. If C does not contain any positive rational, it must be made of only negative irrationals. Since $C \neq \hat{0}$, we can find $r \in \mathbb{Q}^-$ with $r \notin C$, hence C is negative. $\qquad\square$

We want to define the operation of addition for cuts. To find the sum of the cuts $\hat{r}$ and $\hat{s}$, we can just take $\widehat{r+s}$, but since there are other types of cuts, this idea will not suffice.

Definition 11.10. *Given cuts C, D, define their sum by*

$$C + D = \{x + y : x \in C \text{ and } y \in D\}.$$

Theorem 11.11. *If C and D are cuts, then $C + D$ is a cut.*

Proof. Naturally, there are four parts to the proof because there are four separate properties we need to establish.

1. By property 1 of cuts, there is an x in C and there is a y in D. Thus $x + y \in C + D$, so $C + D$ is not empty.

2. Let $u \notin C$ and $v \notin D$. Suppose $x \in C$ and $y \in D$. We have $x < u$ and $y < v$. Thus, by adding the inequalities, $x + y < u + v$. This shows that the rational number $u + v$ is not in $C + D$, because it is not equal to any element of that set.

3. Suppose $z \in C + D$. Then there exist $x \in C$ and $y \in D$ such that $z = x + y$. By property 3 of cuts, there is a $y' \in D$ for which $y < y'$. By adding x to both sides of this inequality we obtain $x + y < x + y'$. Since $x + y' \in C + D$ and $z < x + y'$, we have determined an element of $C + D$ which is larger than z.

4. Consider an arbitrary member of $C + D$. It has the form $x + y$, for some $x \in C$ and $y \in D$. If $r < x + y$, then certainly $r - x < y$. Hence $r - x \in D$ by property 4 of cuts. But $r = x + (r - x)$, so r is an element of $C + D$. $\qquad\square$

Theorem 11.12. *We have*
1. *Addition of cuts is commutative and associative.*
2. *For any cut C, $C + \hat{0} = C$.*

Proof. Part 1 follows from the corresponding properties of rational numbers. For part 2, let $c \in C$ and $z \in \hat{0}$. Then $c + z < c + 0 = c$, hence $C + \hat{0} \subseteq C$. For the other inclusion, let $c \in C$ and choose $d \in C$ with $c < d$. Then $c - d \in \hat{0}$ and $c = d + (c - d) \in C + \hat{0}$. $\square$

Definition 11.13. *If C is a cut, define $-C = \{x \in \mathbb{Q} : x + c < 0 \text{ for all } c \in C\}$.*

We could also define $-C$ as $-C = \{x \in \mathbb{Q} : x < -y \text{ for some } y \notin C\}$. Notice that $-C$ is not the set $\{-c : c \in C\}$.

Lemma 11.14. *Let $r > 0$ be rational and let C be a cut. Then there are $y \in C$ and $z \notin C$ such that $z - y = r$.*

Proof. Let $x \in C$. By the Archimedean property of the rationals, there is a smallest positive integer n such that $x + nr \notin C$. Let $z = x + nr$ and $y = x + (n-1)r$. Then $y \in C$ and $z - y = r$. $\square$

Theorem 11.15. *We have the following:*
1. *If C is a cut, then $-C$ is a cut and $C + (-C) = \hat{0}$.*
2. *If C is a positive cut, then $-C$ is a negative cut.*
3. *If C is a negative cut, then $-C$ is a positive cut.*
4. *$C + D = C + E \Rightarrow D = E$.*

Proof. 1. First we show that $-C$ is a cut, by verifying the four defining properties. If $x > c$ for all $c \in C$, then $-x \in -C$, hence $-C$ is not empty. Given $c \in C$ we have $-c \notin -C$ since $(-c) + c = 0 \geq 0$, hence $-C \neq \mathbb{Q}$. Given $x \in -C$ let us find $y \in -C$ with $x < y$. We have $x + c < 0$ for all $c \in C$. Fixing $c \in C$ arbitrary, let $d \in \mathbb{Q}^+$ with $c + d \in C$. Then $x + d + c < 0$, hence $y = x + d \in -C$ and $x < x + d$. Finally, if $x \in -C$ and $y \in \mathbb{Q}$ with $y < x$, we have $y + c < x + c < 0$ for all $c \in C$, hence $y \in -C$.

By definition, $C + (-C) \subseteq \hat{0}$. If $z \in \hat{0}$, then by Lemma 11.14 there is $c \in C$ with $c - z \notin C$. Then $z - c \in -C$ and $z = c + (z - c)$, hence $\hat{0} \subseteq C + (-C)$ and we have equality.

2. Let $r \in C$ with $r > 0$. Then $-r < 0$ and $-r \notin -C$, hence $-C$ is negative. Parts 3 and 4 are left as an exercise.

$\square$

We now define the operation of multiplication.

Definition 11.16. *If C and D are positive cuts, we define their product by*

$$C \cdot D = \{x \in \mathbb{Q} : x \leq c \cdot d \text{ for some positive } c \in C, d \in D\}.$$

If $C = \hat{0}$ or $D = \hat{0}$, then we define $C \cdot D = \hat{0}$;

if $C \succ \hat{0}$ and $D \prec \hat{0}$, then we define $C \cdot D = -(C \cdot (-D))$;
if $C \prec \hat{0}$ and $D \succ \hat{0}$, then we define $C \cdot D = -((-C) \cdot D)$;
if $C \prec \hat{0}$ and $D \prec \hat{0}$, then we define $C \cdot D = (-C) \cdot (-D)$.

Theorem 11.17. *We have*

1. *The product of two positive cuts is a positive cut.*
2. *Multiplication of cuts is commutative and associative.*
3. *Multiplication is distributive over addition of cuts.*
4. *If C is a cut, then $C \cdot \hat{1} = C$.*

Proof. 1. To show that $C \cdot D$ is a cut, first observe that $C \cdot D$ is not empty and not equal to $\mathbb{Q}$. Also, given $x \in C \cdot D$, there are $c \in C$ and $d \in D$ positive with $x \leq c \cdot d$. Let $c' > 0$ with $c + c' \in C$. Then $x + c'd \leq (c + c') \cdot d$, hence $x + c'd \in C \cdot D$ and $x < x + c'd$. The fourth property is obvious. To show that $C \cdot D$ is positive, observe that for $c \in C$ and $d \in D$ positive, $c \cdot d$ is positive and $c \cdot d \in C \cdot D$.

2. For positive cuts, $C \cdot D = D \cdot C$ and $C \cdot (D \cdot E) = (C \cdot D) \cdot E$ by definition and the properties of multiplication of rational numbers. If for example $C \succ \hat{0}$, $D \prec \hat{0}$ and $E \prec \hat{0}$, then $C \cdot D = -(C \cdot (-D))$ and $D \cdot C = -((-D) \cdot C)$, hence $C \cdot D = D \cdot C$. Also $C \cdot (D \cdot E) = C \cdot ((-D) \cdot (-E)) = (C \cdot (-D)) \cdot (-E)$ and $(C \cdot D) \cdot E = (-(C \cdot (-D))) \cdot E = (C \cdot (-D)) \cdot (-E)$, hence $C \cdot (D \cdot E) = (C \cdot D) \cdot E$. The other cases are treated similarly.

3. Let us show that $C \cdot (D + E) = C \cdot D + C \cdot E$ for positive cuts. Let $x \in C \cdot (D + E)$. Then $x \leq c \cdot (d + e)$ for some positive $c \in C, d \in D, e \in E$. We get $x \leq c \cdot d + c \cdot e$ with $c \cdot d \in C \cdot D, c \cdot e \in C \cdot E$, hence $x \in C \cdot D + C \cdot E$. Conversely, let $x \in C \cdot D + C \cdot E$. Then $x = y + z$ with $y \in C \cdot D, z \in C \cdot E$, so $y \leq c_1 \cdot d$ and $z \leq c_2 \cdot e$ for positive $c_1, c_2 \in C, d \in D, e \in E$. Let $c = \max(c_1, c_2)$. Then $x \leq c \cdot d + c \cdot e = c \cdot (d + e)$, hence $x \in C \cdot (D + E)$.

If for example $C \succ \hat{0}, D \prec \hat{0}$ and $D + E \succ \hat{0}$, then $E = (D + E) + (-D)$. We get $C \cdot E = C \cdot (D + E) + C \cdot (-D) = C \cdot (D + E) - C \cdot D$, hence $C \cdot D + C \cdot E = C \cdot (D + E)$. The other cases are similar.

4. It suffices to consider the case $C \succ \hat{0}$. Let $x \in C \cdot \hat{1}$. Then $x \leq y \cdot z$ for some positive $y \in C$ and $z \in \hat{1}$. Since $z < 1$, we get $y \cdot z < y$ and $x < y$, so $x \in C$ by the properties of cuts. Conversely, let $x \in C$. Consider $y \in C$ positive with $y > x$. Then $x = y \cdot \frac{x}{y}$ with $\frac{x}{y} \in \hat{1}$, hence $x \in C \cdot \hat{1}$. $\square$

Definition 11.18. *If C is a positive cut, define*

$$C^{-1} = \{y \in \mathbb{Q} : y < \frac{1}{x} \text{ for some } x \notin C\}.$$

Lemma 11.19. *If C is a positive cut, then C^{-1} is a positive cut.*

Proof. To show that C^{-1} is a cut, we follow the four steps:

1. By definition, there is an $x \notin C$. Since $C \succ \hat{0}$, we have $x > 0$. Since $x + 1 > x$, we have $\frac{1}{x+1} < \frac{1}{x}$, so $\frac{1}{x+1} \in C^{-1}$ and $\frac{1}{x+1} > 0$. Thus $C^{-1} \neq \emptyset$ and it contains a positive element.

2. Let $y \in C$ such that $y > 0$. Then $\frac{1}{y} \notin C^{-1}$. Indeed, if $\frac{1}{y} \in C^{-1}$, then $\frac{1}{y} < \frac{1}{z}$ for some $z \notin C$. But then $z < y$, and hence $z \in C$, a contradiction.

3. Let $x \in C^{-1}$. Then $x < \frac{1}{y}$ for some $y \notin C$. Take $x' \in \mathbb{Q}$ with $x < x' < \frac{1}{y}$, for example, their average. Since, in particular, $x' < \frac{1}{y}$, we have $x' \in C^{-1}$.

4. If $x \in C^{-1}$ and $r < x$, then $x < \frac{1}{y}$ for some $y \notin C$. By transitivity, $r < \frac{1}{y}$, so $r \in C^{-1}$.

Since C^{-1} contains the positive element $\frac{1}{x+1}$, it is a positive cut. $\qquad\square$

Theorem 11.20. *If C is a positive cut, then $C \cdot C^{-1} = \hat{1}$.*

Proof. Since both $C \cdot C^{-1}$ and $\hat{1}$ contain all rationals less or equal to 0, it suffices to show that they contain the same positive rationals. Suppose $t \in C \cdot C^{-1}$ and $t > 0$. Then for some positive $x \in C$ and $y \in C^{-1}$ we have $t \leq x \cdot y$. Since $y < 1/z$ for some $z \notin C$, we get $z > x$. Taking reciprocals yields $1/z < 1/x$, which implies $y < 1/x$. Hence $x \cdot y < 1$, so, since $t \leq x \cdot y$, clearly $t \in \hat{1}$. We get $C \cdot C^{-1} \subseteq \hat{1}$.

On the other hand, if $x \in \hat{1}$, with $x > 0$, we can select an $a \in C$ such that $a > 0$ and by Lemma 11.14 (with $r = a(1-x)$) we can find $y \in C$ and $z \notin C$ such that $z - y = a(1-x)$. Since $a \in C$, we have $a < z$, so $a(1-x) < z(1-x)$ since $1 - x > 0$. We obtain $z - y < z(1-x)$, hence $z - y < z - zx$, $-y < -zx$, $y > zx$, and therefore $y/x > z$. Thus $x/y < 1/z$, so $x/y \in C^{-1}$. Since $x = y \cdot \frac{x}{y}$ with $y \in C$, we get $x \in C \cdot C^{-1}$, so $\hat{1} \subseteq C \cdot C^{-1}$. $\qquad\square$

Corollary 11.21. *If C, D, and E are positive cuts, and $C \cdot D = C \cdot E$, then $D = E$.*

If C is a positive cut, then $(C^{-1})^{-1} = C$.

If D is a positive cut and $C \succ \hat{1}$, then $C \cdot D \succ D$.

Proof. We prove the last part and leave the rest as an exercise. Suppose $C \cdot D$ is not greater than D. By trichotomy, either $C \cdot D = D$ or $C \cdot D \prec D$.

Case 1. If $C \cdot D = D$, then $C \cdot D \cdot D^{-1} = D \cdot D^{-1}$. This implies that $C \cdot \hat{1} = \hat{1}$, hence $C = \hat{1}$, a contradiction.

Case 2. Suppose $C \cdot D \prec D$. Then there is an $x \in D$ such that $x \notin C \cdot D$. Clearly $x > 0$, since $C \cdot D$ is positive. But since $1 \in C$, we have $1 \cdot x \in C \cdot D$, a contradiction.

Here is another proof of the same statement. Suppose $x \in D$. Then $1 \cdot x = x \in C \cdot D$, because $1 \in C$. Thus $D \subseteq C \cdot D$, so $D \preceq C \cdot D$. If equality held, we could multiply both sides by D^{-1} and obtain $\hat{1} = C$, an obvious contradiction. $\qquad\square$

Lemma 11.22. *If $x > 1$ and $x \in C$, then $C^{-1} \prec \hat{1}$.*

Proof. Since $x \in C$ and $x > 0$, we have $1/x \notin C^{-1}$. But since $x > 1$, it follows that $1/x < 1$, so $1/x \in \hat{1}$. By definition, $C^{-1} \prec \hat{1}$. $\qquad\square$

Definition 11.23. *The absolute value of a cut is defined as* $|C| = \max(C, -C)$.

Definition 11.24. *If C is a negative cut, then we define $C^{-1} = -(-C)^{-1}$.*

Theorem 11.25. *1. Given $C \neq \hat{0}$ we have $C \cdot C^{-1} = \hat{1}$.*
2. If $A + C = B$ and C is a positive cut, then $A \prec B$.
3. $A \prec B \Leftrightarrow A + C \prec B + C$.
4. If $C \prec D$ and E is positive, then $C \cdot E \prec D \cdot E$.

Proof. 4. Let $p \in D \setminus C$ and let $q \in D$ with $q > p$. For $c \in E$ positive, set $c_n = c\dfrac{q^{n-1}}{p^{n-1}}$. Consider m the positive integer such that $c_m \in E$ and $c_{m+1} \notin E$. We have $pc_{m+1} = qc_m$. Moreover, $pc_{m+1} > uv$ for all $u \in C$ and $v \in E$ and $pc_{m+1} \notin C \cdot E$. We found $qc_m \in D \cdot E \setminus C \cdot E$, hence $C \cdot E \prec D \cdot E$.

We leave the other parts as exercise. $\qquad \square$

Definition 11.26. *Each Dedekind cut will be called a* real number. *The set of real numbers will be denoted by $\mathbb{R}$. The order relation $\preceq$ on $\mathbb{R}$ will be also denoted by $\leq$.*

Theorem 11.27. *If $A = \{\hat{r} : r \in \mathbb{Q}\} \subseteq \mathbb{R}$, then $\mathbb{Q}$ is isomorphic with A. In particular the set of rational numbers $\mathbb{Q}$ can be viewed as a subset of $\mathbb{R}$.*

Proof. Define $f : \mathbb{Q} \to A, f(r) = \hat{r}$. Then f is one-to-one and onto, $f(r+s) = f(r) + f(s)$, $f(rs) = f(r) \cdot f(s)$ and $r \leq s \Rightarrow f(r) \leq f(s)$. $\qquad \square$

As a result, we shall feel free to make the same simplifying conventions as before. In other words r and $\hat{r}$ will be regarded as the same for each rational number r.

Theorem 11.28. *Suppose that C_i is a cut for each $i \in I \neq \emptyset$. If $C = \bigcup_{i \in I} C_i$ and $C \neq \mathbb{Q}$, then C is a cut.*

Proof. Clearly C is not empty and $C \neq \mathbb{Q}$ by hypothesis. Let $x \in C$. There is $i_0 \in I$ with $x \in C_{i_0}$. Since C_{i_0} is a cut, there is $y \in C_{i_0}$ with $x < y$. But then $y \in C$ and $x < y$. If $r < x$, then $r \in C_{i_0}$, hence $r \in C$. $\qquad \square$

Suppose $S \subseteq \mathbb{R}$ is a set of real numbers. Recall that a real number a is an *upper bound* for S iff $a \geq x$ for all $x \in S$; a real number a is a *lower bound* for S iff $a \leq x$ for all $x \in S$.

The number a is a *least upper bound* or *supremum* for S iff a is an upper bound for S and $t < a \Rightarrow \exists x \in S$ with $t < x$. Using quantifiers, $a = \sup S$ if and only if

$$\forall s \in S, s \leq a \text{ and } \forall \varepsilon > 0 \ \exists x \in S \text{ such that } a - \varepsilon < x.$$

Indeed, $a - \varepsilon$ is no longer an upper bound for S.

The number a is a *greatest lower bound* or *infimum* for S iff a is a lower bound for S and $t > a \Rightarrow \exists x \in S$ with $x < t$.

Corollary 11.29. *We have*

1. If S is a nonempty subset of $\mathbb{R}$ that has an upper bound, then S has a least upper bound.

2. If S is a nonempty subset of $\mathbb{R}$ that has a lower bound, then S has a greatest lower bound.

Proof. Let $T = \bigcup_{C \in S} C$. Then T is a cut and it is an upper bound for S. To show that T is the least upper bound, let $U \prec T$, hence there is $q \in T \setminus U$. By definition, there is $C \in S$ with $q \in C$, hence $C \succ U$ and U cannot be an upper bound for S. Hence T is the least upper bound of S.

2. We use $\text{glb}S = -\text{lub}(-S)$, where here $-S = \{-s : s \in S\}$. We leave the proof as an exercise. $\qquad\square$

The above property of $\mathbb{R}$ is called the *least upper bound property*. The set $\mathbb{Q}$ does not have this property. Indeed, the set $\{x \in \mathbb{Q}^+ : x^2 < 2\}$ has no least upper bound in $\mathbb{Q}$, since $\sqrt{2}$ is irrational.

We summarize the properties of the set of real numbers in the following theorem.

Theorem 11.30. *The set $\mathbb{R}$ with the addition and multiplication operations and with the order relation $\leq$ contains a copy of $\mathbb{Q}$ and satisfies these properties:*

1. $\mathbb{R}$ is closed under addition and multiplication.

2. Addition and multiplication are commutative and associative.

3. $\mathbb{R}$ contains 0 and $x + 0 = x$ $\forall x \in \mathbb{R}$.

4. Each $x \in \mathbb{R}$ has an opposite $-x$ such that $x + (-x) = 0$.

5. $\mathbb{R}$ contains 1 such that $1 \neq 0$ and $x \cdot 1 = x$ $\forall x \in \mathbb{R}$.

6. Any $x \in \mathbb{R} \setminus \{0\}$ has an inverse x^{-1} such that $x \cdot x^{-1} = 1$.

7. Multiplication is distributive over addition.

8. We have $(x > 0) \wedge (y > 0) \Rightarrow (x + y > 0) \wedge (xy > 0)$ and for any $x \in \mathbb{R}$ exactly one is true: $x > 0, -x > 0$ or $x = 0$.

9. Any nonempty subset of $\mathbb{R}$ which has an upper bound has a least upper bound.

We say that $\mathbb{R}$ is an ordered field with the least upper bound property, or a *complete ordered field*.

Theorem 11.31. *The set of real numbers also has these properties:*

1. (Archimedean property) If $x, y \in \mathbb{R}$ and $x > 0$, then there is a positive integer n such that $nx > y$.

2. If $x, y \in \mathbb{R}$ and $x < y$, then there is $r \in \mathbb{Q}$ such that $x < r < y$.

Proof. 1. Let $A = \{nx : n \in \mathbb{P}\}$. If the assertion is false, then y would be an upper bound for A. Let $\alpha = \sup A$. Since $x > 0$, we get $\alpha - x < \alpha$ and $\alpha - x$ is no longer an upper bound for A. Hence there is m with $\alpha - x < mx$, so $\alpha < (m + 1)x \in A$, a contradiction.

2. Since $x < y$, we have $y - x > 0$ and from 1 we can find $n \in \mathbb{P}$ such that $n(y - x) > 1$. Similarly, we obtain positive integers m_1 and m_2 such that $m_1 > nx$ and $m_2 > -nx$. We get $-m_2 < nx < m_1$ and we can find an integer m such that $m - 1 \le nx < m$. Combining the inequalities we get $nx < m \le 1 + nx < ny$. Dividing by n we get $x < \dfrac{m}{n} < y$. $\square$

It follows that $\mathbb{R}$ is a complete ordered field with the Archimedean property.

Theorem 11.32. *For any real $x > 0$ and any positive integer n there is a unique real $y > 0$ such that $y^n = x$. We write $y = \sqrt[n]{x}$.*

Proof. Let A be the set of all positive real numbers t such that $t^n < x$. To prove that A is not empty, consider $s = \dfrac{x}{x + 1}$. Then $0 < s < 1$ and $s^n < s < x$, hence $s \in A$. It is easy to see that $1 + x$ is an upper bound for A. We conclude that $y = \sup A$ exists. Let us prove that $y^n = x$.

Assume $y^n < x$. Choose h such that $0 < h < 1$ and $h < \dfrac{x - y^n}{n(y+1)^{n-1}}$. Then

$$(y + h)^n - y^n < hn(y + h)^{n-1} < hn(y + 1)^{n-1} < x - y^n,$$

hence $(y + h)^n < x$ and $y + h \in A$, a contradiction. Similarly, the assumption $y^n > x$ leads to a contradiction. We conclude that $y^n = x$. $\square$

It can be proved that for $b > 1$ and $x \in \mathbb{R}$ there is a real number b^x such that for $x, y \in \mathbb{R}$ we have $b^{x+y} = b^x b^y$. Also, given $y > 0$ there is a unique real number x such that $b^x = y$. This is called the logarithm of y in base b, denoted $\log_b y$.

Exercise 11.33. *There are irrational numbers x, y such that x^y is rational.*

Solution. Indeed, consider $\sqrt{2}^{\sqrt{2}}$. If this is rational, take $x = y = \sqrt{2}$. If not, then

$$(\sqrt{2}^{\sqrt{2}})^{\sqrt{2}} = \sqrt{2}^2 = 2,$$

so we can take $x = \sqrt{2}^{\sqrt{2}}$ and $y = \sqrt{2}$.

11.2   The Cauchy sequences approach

The set of real numbers can be constructed using equivalence classes of certain sequences of rational numbers. You may think of the real numbers as being limits of sequences of rational numbers. This is an analytic process and the proofs might be more involved, but the construction illustrates the more general concept of *completion*, which you may encounter later in functional analysis and other branches of mathematics.

Recall that a sequence in a set S is a function $x : \mathbb{N} \to S$. The element $x(n) \in S$ is denoted x_n. We also write a sequence as $x = (x_n)_{n \geq 0}$ or just (x_n). Sometimes a sequence (x_n) is defined only for $n \geq k$ for $k \in \mathbb{N}$. You have studied sequences and limits in calculus. We will prove that the limit of a sequence of rational numbers is not necessarily a rational number. The real numbers, containing both rational and irrational numbers, are filling those gaps.

Definition 11.34. *A sequence* $x = (x_n)$ *of rational numbers is a* Cauchy sequence *if for every positive rational* ε *there is a positive integer* N *such that for every* $m, n \geq N$ *we have* $|x_n - x_m| < \varepsilon$.

Note that this definition retains the flavor of the definition of the limit. Recall that $L = \lim\limits_{n \to \infty} x_n$ if for any $\varepsilon > 0$ there is N such that $|x_n - L| < \varepsilon$ for all $n \geq N$. However, we will see that a Cauchy sequence in $\mathbb{Q}$ is not necessarily convergent in $\mathbb{Q}$.

Example 11.35. *Obviously, a constant sequence* $r, r, r, \ldots$ *is a Cauchy sequence. For any* $r \in \mathbb{Q}$, *denote by* $\hat{r}$ *the constant sequence* $r, r, r, \ldots$

Example 11.36. *Let* $x_n = \dfrac{n+1}{n}$, $n \geq 1$. *Then* (x_n) *is a Cauchy sequence. Indeed,*

$$|x_n - x_m| = \left| \frac{m-n}{mn} \right| = \left| \frac{1}{n} - \frac{1}{m} \right| < \frac{1}{\min(m,n)}.$$

Given $\varepsilon > 0$, *take* $N = \lfloor 1/\varepsilon \rfloor + 1$. *For* $m, n \geq N$ *we have*

$$|x_n - x_m| < \frac{1}{N} < \varepsilon.$$

Example 11.37. *The sequence such that* $x_1 = 0$, $x_2 = 1$, *and* $x_n = \frac{1}{2}(x_{n-1} + x_{n-2})$ *for* $n \geq 3$ *is a Cauchy sequence.*

Proof. First, let's prove by induction that

$$x_{n+1} - x_n = \frac{(-1)^{n-1}}{2^{n-1}}, n \geq 1.$$

For $n = 1$, $x_2 - x_1 = 1 - 0 = 1 = \frac{(-1)^0}{2^0}$. Suppose $x_{k+1} - x_k = \frac{(-1)^{k-1}}{2^{k-1}}$ for a fixed $k \geq 1$. Then

$$x_{k+2} - x_{k+1} = \frac{1}{2}(x_{k+1} + x_k) - x_{k+1} = -\frac{1}{2}(x_{k+1} - x_k) = \frac{(-1)^k}{2^k}.$$

From the recurrence formula, it is clear that for $m > n$, x_m lies between x_n and x_{n+1}. Given a positive ε, choose N such that $2^{N-1} < 1/\varepsilon$. Then for all $m, n \geq N$,

$$|x_m - x_n| \leq |x_{n+1} - x_n| = \frac{1}{2^{n-1}} < \frac{1}{2^{N-1}} < \varepsilon.$$

$\square$

Example 11.38. *Let $x_1 = \dfrac{1}{2}, x_{n+1} = \dfrac{1}{2+x_n}, n \geq 1$. Then (x_n) is a Cauchy sequence of rational numbers.*

Proof. It is clear that $x_n \in \mathbb{Q}$ and $x_n > 0$ for all n. We compute

$$x_{n+1} - x_{n+2} = \frac{1}{2+x_n} - \frac{1}{2+x_{n+1}} = \frac{x_{n+1} - x_n}{(2+x_n)(2+x_{n+1})}.$$

Since $x_n > 0$, we get $(2+x_n)(2+x_{n+1}) > 4$ and therefore

$$|x_{n+1} - x_{n+2}| < \frac{1}{4}|x_n - x_{n+1}|$$

for all $n \geq 1$. By induction we can prove that $|x_n - x_{n+1}| < \frac{1}{4^{n-1}}$. Indeed,

$$|x_1 - x_2| = \left|\frac{1}{2} - \frac{2}{5}\right| = \frac{1}{10} < 1.$$

Assume $|x_k - x_{k+1}| < \frac{1}{4^{k-1}}$. Then

$$|x_{k+1} - x_{k+2}| < \frac{1}{4}|x_k - x_{k+1}| < \frac{1}{4^k}.$$

To prove that (x_n) is Cauchy, fix $\varepsilon > 0$ rational and choose N such that $\dfrac{1}{4^{N-1}} < \dfrac{3}{4}\varepsilon$. For $m > n \geq N$ we have

$$|x_m - x_n| \leq |x_m - x_{m-1}| + |x_{m-1} - x_{m-2}| + \cdots + |x_{n+1} - x_n| <$$

$$< \frac{1}{4^{m-2}} + \frac{1}{4^{m-3}} + \cdots + \frac{1}{4^{n-1}} =$$

$$\frac{1}{4^{n-1}}\left(1 + \frac{1}{4} + \cdots + \frac{1}{4^{m-n-1}}\right) = \frac{1}{4^{n-1}}\frac{1 - \frac{1}{4^{m-n}}}{1 - \frac{1}{4}} < \frac{4}{3}\frac{1}{4^{n-1}} < \frac{4}{3}\frac{1}{4^{N-1}} < \varepsilon.$$

For $n > m \geq N$ we similarly obtain $|x_m - x_n| < \varepsilon$, hence (x_n) is Cauchy. It can be proved that $\lim_{n \to \infty} x_n = \sqrt{2} - 1$, an irrational number.

$\square$

Definition 11.39. *We define addition and multiplication of sequences of rational numbers $x = (x_n)$ and $y = (y_n)$ by*

$$x + y = u \text{ where } u_n = x_n + y_n; \quad xy = v \text{ where } v_n = x_n y_n.$$

Clearly, if x and y are sequences of rational numbers, then so are $x + y$ and xy.

Theorem 11.40. *If x and y are Cauchy sequences of rational numbers, then $x + y$ and xy are Cauchy sequences of rational numbers.*

Proof. Let $\varepsilon > 0$. There exist N_1 and N_2 such that for all $m, n \geq N_1$ we have $|x_n - x_m| < \varepsilon/2$ and for all $m, n \geq N_2$ we have $|y_n - y_m| < \varepsilon/2$. Let $N = \max\{N_1, N_2\}$. For $m, n \geq N$ we have

$$|(x_n + y_n) - (x_m + y_m)| = |(x_n - x_m) + (y_n - y_m)| \leq$$
$$\leq |x_n - x_m| + |y_n - y_m| < \varepsilon/2 + \varepsilon/2 = \varepsilon.$$

For the product, we need to show first that a Cauchy sequence x is bounded, i.e., there exists a positive rational number α such that $|x_n| \leq \alpha$ for all n. Indeed, for $\varepsilon = 1$, there is N such that $|x_n - x_m| < 1$ for all $m, n \geq N$. Let

$$\alpha = \max\{|x_1|, |x_2|, \cdots, |x_N|\} + 1.$$

Clearly, for $n \leq N$ we have $|x_n| \leq \alpha$. For $n > N$ we have $|x_n - x_N| < 1$, hence $|x_n| \leq |x_n - x_N| + |x_N| \leq 1 + |x_N|$ and for all n we get $|x_n| \leq \alpha$.

Given $\varepsilon > 0$, there are α_1 and α_2 positive such that $|x_n| \leq \alpha_1$ and $|y_n| \leq \alpha_2$ for all n. There are integers N_1 and N_2 such that for all $m, n \geq N_1$ we have $|x_n - x_m| < \varepsilon/(2\alpha_2)$ and for all $m, n > N_2$ we have $|y_n - y_m| < \varepsilon/(2\alpha_1)$. Let $N = \max\{N_1, N_2\}$. For all $m, n \geq N$ we have

$$|x_n y_n - x_m y_m| = |x_n y_n - x_m y_n + x_m y_n - x_m y_m| \leq$$
$$\leq |y_n||x_n - x_m| + |x_m||y_n - y_m| <$$
$$< \alpha_2 \frac{\varepsilon}{2\alpha_2} + \alpha_1 \frac{\varepsilon}{2\alpha_1} = \varepsilon.$$

$\square$

Theorem 11.41. *The addition and multiplication of Cauchy sequences are commutative and associative. Multiplication is distributive over addition. Each sequence x has an opposite $-x$, the sequence $\hat{0}$ satisfies $x + \hat{0} = x$, and the sequence $\hat{1}$ satisfies $x \cdot \hat{1} = x$.*

Proof. These properties follow directly from the definition and from the corresponding properties of the rational numbers. $\square$

Definition 11.42. *We define the relation $\sim$ on the set of Cauchy sequences of rational numbers by $x \sim y$ iff for any positive rational number ε there is an integer N such that $|x_n - y_n| < \varepsilon$ for all $n \geq N$.*

Example 11.43. *Let $x_n = \dfrac{n+1}{n}$ and $y_n = 1$ for all n. Clearly, $x \sim y$ since* $x_n - y_n = \dfrac{1}{n}$.

Theorem 11.44. *The relation $\sim$ is an equivalence relation.*

Proof. Indeed, it is reflexive since $x \sim x$ and it is symmetric since $x \sim y \Rightarrow y \sim x$. Let $x \sim y$ and $y \sim z$. Given $\varepsilon > 0$, there are N_1, N_2 such that $|x_n - y_n| < \varepsilon/2$ for $n \geq N_1$ and $|y_n - z_n| < \varepsilon/2$ for $n \geq N_2$. Let $N = \max\{N_1, N_2\}$. Then $|x_n - z_n| \leq |x_n - y_n| + |y_n - z_n| \leq \varepsilon/2 + \varepsilon/2 = \varepsilon$ for all $n \geq N$, hence $\sim$ is transitive. $\square$

Definition 11.45. *A Cauchy sequence $x = (x_n)$ is called* positive *iff there is $\varepsilon > 0$ and an integer N such that $x_n > \varepsilon$ for all $n \geq N$.*

Lemma 11.46. *If x, y, u, v are Cauchy sequences in $\mathbb{Q}$ such that $x \sim u$ and $y \sim v$, then $x + y \sim u + v$ and $xy \sim uv$. Moreover, if x is positive, then u is positive.*

Proof. Given $\varepsilon > 0$, choose N such that $|x_n - u_n| < \varepsilon/2$ and $|y_n - v_n| < \varepsilon/2$ for all $n \geq N$. Then for all $n \geq N$ we have

$$|(x_n + y_n) - (u_n + v_n)| \leq |x_n - u_n| + |y_n - v_n| < \varepsilon/2 + \varepsilon/2 = \varepsilon,$$

hence $x + y \sim u + v$.

Since y, u are bounded, we can find $\alpha > 0$ such that $|y_n|, |u_n| \leq \alpha$ for all n. Given $\varepsilon > 0$, choose N such that $|x_n - u_n| < \frac{\varepsilon}{2\alpha}$ and $|y_n - v_n| < \frac{\varepsilon}{2\alpha}$ for all $n \geq N$. Then for $n \geq N$ we have

$$|x_n y_n - u_n v_n| \leq |x_n y_n - u_n y_n| + |u_n y_n - u_n v_n| = |y_n||x_n - u_n| + |u_n||y_n - v_n| <$$

$$< \alpha \frac{\varepsilon}{2\alpha} + \alpha \frac{\varepsilon}{2\alpha} = \varepsilon,$$

hence $xy \sim uv$.

Assume now that x is positive and $x \sim u$. There is $\varepsilon > 0$ and N_0 such that $x_n > \varepsilon$ for all $n \geq N_0$. For this $\varepsilon > 0$ there is $N \geq N_0$ such that $|x_n - u_n| < \varepsilon/2$ for all $n \geq N$. Then for $n \geq N$ we have $u_n > x_n - \varepsilon/2 > \varepsilon/2$, hence u is positive. $\qquad\square$

Theorem 11.47. *We have*

1. The sum and the product of two positive Cauchy sequences are positive.

2. If x is any Cauchy sequence, then exactly one of the following holds true: x is positive, $x \sim \hat{0}$, or $-x$ is positive.

3. If the Cauchy sequence x is not equivalent to $\hat{0}$, there is a Cauchy sequence z with $xz \sim \hat{1}$.

Proof. 1. Let x, y be positive Cauchy sequences. There are $\varepsilon_i > 0$ and integers N_i for $i = 1, 2$ such that $x_n > \varepsilon_1$ for $n \geq N_1$ and $y_n > \varepsilon_2$ for $n \geq N_2$. Then $x_n + y_n > \varepsilon_1 + \varepsilon_2$ and $x_n y_n > \varepsilon_1 \varepsilon_2$ for $n \geq \max\{N_1, N_2\}$.

2. If $x \not\sim \hat{0}$, then there is $\varepsilon > 0$ such that for any integer N there is $n > N$ such that $|x_n| > \varepsilon$. Since x is Cauchy, there is N_1 such that $|x_n - x_m| < \varepsilon/2$ for $m, n \geq N_1$. In particular, for $N = N_1$ there is $p > N_1$ such that $|x_p| > \varepsilon$. To make a choice, let $x_p > 0$, hence $x_p > \varepsilon$. We have $|x_n - x_p| < \varepsilon/2$ for all $n \geq p$, hence $x_n \geq x_p - \varepsilon/2 > \varepsilon - \varepsilon/2 = \varepsilon/2$ and x is positive. If $x_p < 0$ a similar argument shows that $-x$ is positive.

3. Since $x \not\sim \hat{0}$, from part 2 we get that there is a positive rational ε' and an integer N such that $|x_n| \geq \varepsilon'$ for all $n \geq N$. Consider x' such that $x'_n = \varepsilon'$ for $n < N$ and $x'_n = x_n$ for $n \geq N$. Since $x'_n = x_n$ for $n \geq N$, and x is Cauchy, it follows that x' is also Cauchy and by definition we have $|x'_n| \geq \varepsilon'$ for all

n. In particular, $x'_n \neq 0$ for all n. Define the sequence z such that $z_n = \frac{1}{x'_n}$. Given $\varepsilon > 0$, consider N such that $|x'_m - x'_n| < \varepsilon\varepsilon'^2$ for all $m, n \geq N$. Such an N exists since x' is Cauchy. Since $|x'_n| \geq \varepsilon'$ for all n, we have $\dfrac{1}{|x'_m x'_n|} \leq \dfrac{1}{\varepsilon'^2}$ for all m, n. We obtain

$$|z_m - z_n| = \left| \frac{1}{x'_m} - \frac{1}{x'_n} \right| = \frac{|x'_m - x'_n|}{|x'_m x'_n|} < \frac{\varepsilon\varepsilon'^2}{\varepsilon'^2} = \varepsilon$$

for all $m, n \geq N$, therefore z is Cauchy. It is clear that $zx' \sim \hat{1}$, and since $x' \sim x$ we conclude that $zx \sim \hat{1}$. $\qquad\square$

Definition 11.48. *A real number is an equivalence class $[x]$ of Cauchy sequences of rational numbers. We say that $[x]$ is positive iff it contains a positive Cauchy sequence. Define addition and multiplication by*

$$[x] + [y] = [x + y], \quad [x] \cdot [y] = [xy].$$

Define $[x] \prec [y]$ iff $[y] - [x]$ is positive.

Note that there is a one-to-one function $f : \mathbb{Q} \to \mathbb{R}$ such that $f(r) = [\hat{r}]$, which preserves the order and the operations. This allows us to make an identification of $\mathbb{Q}$ with a subset of $\mathbb{R}$. From now on, we will use x instead of $[x]$, $x < y$ instead of $[x] \prec [y]$, 0 instead of $\hat{0}$, and 1 instead of $\hat{1}$. As usual, the absolute value function is defined as $|x| = \max(x, -x)$.

We summarize the properties of real numbers so far. The set $\mathbb{R}$ with the operations of addition and multiplication, 0, 1 and with strict order $<$ satisfy the following:
  1) $x + (y + z) = (x + y) + z$.
  2) $x + y = y + x$.
  3) $0 + x = x$.
  4) For each x there is z with $x + z = 0$.
  5) $x(yz) = (xy)z$.
  6) $xy = yx$.
  7) $1 \cdot x = x$.
  8) If $x \neq 0$, there exists z such that $xz = 1$.
  9) $x(y + z) = xy + xz$.
  10) $x, y > 0$ implies $x + y, xy > 0$.
  11) Exactly one of $x > 0$, $x = 0$, $-x > 0$ holds.

Theorem 11.49. *Between any two distinct real numbers there is a rational number.*

Proof. Let $x < y$, and let a, b be Cauchy sequences of rational numbers representing x, y. Since $x < y$, we can find a rational ε and an integer N_1 such that $b_n - a_n > 4\varepsilon$ for all $n \geq N_1$. Since a, b are Cauchy sequences, there are N_2, N_3 such that $|a_n - a_m| < \varepsilon$ for $m, n \geq N_2$ and $|b_n - b_m| < \varepsilon$ for $m, n \geq N_3$. Let

$N = \max\{N_1, N_2, N_3\} + 1$ and let $s = 3\varepsilon/2$. Consider $z = a_N + s \in \mathbb{Q}$. We claim that $x < z < y$. Indeed, $a_n - a_N < \varepsilon$ for every $n \geq N$ and therefore $z - a_n = (a_N + s) - a_n > \varepsilon/2 > 0$, so $z - x > 0$. Similarly, $y - z > 0$. $\square$

Theorem 11.50. *(Archimedean property). If x, y are positive real numbers, then there exists a positive integer n such that $nx > y$.*

Proof. Let b be a representative of y. Since Cauchy sequences of rational numbers are bounded, there is $\beta > 0$ rational such that $b_m \leq \beta$ for all m. Let $d = \hat{\beta}$. We have $y \leq d$. Since $x > 0$ we can find rational numbers s, t such that $0 < s < x$ and $y < t < d$. From the Archimedean property for rational numbers, there is a positive integer n such that $ns > t$. We get $nx > ns > t > y$. $\square$

Theorem 11.51. *A nonempty set of real numbers which has an upper bound has a least upper bound.*

Proof. Let A be a set satisfying the hypothesis. Inductively we construct a sequence of rational numbers (b_n) such that b_n is an upper bound for A but $b_n - \frac{1}{2^n}$ is not. Indeed, let b_0 be an integer such that b_0 is an upper bound for A while $b_0 - 1$ is not. For $n \geq 1$ we define

$$
b_n = \begin{cases} b_{n-1} - \frac{1}{2^n} & \text{if } b_{n-1} - \frac{1}{2^n} \text{ is an upper bound} \\[2mm] b_{n-1} & \text{if } b_{n-1} - \frac{1}{2^n} \text{ is not an upper bound.} \end{cases}
$$

For $m \geq n$ we have $b_n - \frac{1}{2^n} < b_m \leq b_n$, hence $|b_n - b_m| < \frac{1}{2^n}$. We conclude that for any fixed N and $m, n \geq N$ we have $|b_n - b_m| < \frac{1}{2^N}$, therefore $b = (b_n)$ is a Cauchy sequence of rational numbers. Consider $u = [b] \in \mathbb{R}$. We claim that $u = \mathrm{lub}A$. It is easy to prove that $b_n - \frac{1}{2^n} \leq u \leq b_n$ (exercise!). If there is $a \in A$ with $a > u$, then we can find n with $\frac{1}{2^n} < a - u$ which gives $b_n < a$, a contradiction. This proves that u is an upper bound for A. If v is an upper bound with $v < u$, choose n such that $\frac{1}{2^n} < u - v$. Since $b_n - \frac{1}{2^n}$ is not an upper bound, there is $a \in A$ such that $b_n - \frac{1}{2^n} < a$, which gives $b_n - \frac{1}{2^n} < v$. Adding this with $\frac{1}{2^n} < u - v$, we obtain $b_n < u$, a contradiction. $\square$

Corollary 11.52. *A nonempty subset of real numbers which has a lower bound has a greatest lower bound.*

Definition 11.53. *A sequence $x = (x_n)$ of real numbers is Cauchy if for any positive real number ε there exists a positive integer N such that $|x_m - x_n| < \varepsilon$ for all $m, n \geq N$.*

  A real number L is a limit of a sequence $x = (x_n)$ of real numbers if for any $\varepsilon > 0$ there is N such that $|x_n - L| < \varepsilon$ for all $n \geq N$. We write $L = \lim\limits_{n \to \infty} x_n$ or $x_n \to L$ and we say that $x = (x_n)$ is convergent to L.

Example 11.54. *Let $x_1 = 1, x_{n+1} = \dfrac{1}{2}\left(x_n + \dfrac{\sqrt{2}}{x_n}\right)$. Then x is a Cauchy sequence of real numbers. In fact, $x_n \to \sqrt[4]{2}$.*

Proof. We have

$$x_{n+1} - x_n = \frac{1}{2}x_n + \frac{\sqrt{2}}{2x_n} - x_n = \frac{\sqrt{2} - x_n^2}{2x_n}.$$

By induction we will prove that $x_n \geq \sqrt[4]{2}$ for all $n \geq 2$, hence $x_{n+1} \leq x_n$ for $n \geq 2$. Indeed, $x_2 = \dfrac{1 + \sqrt{2}}{2} \geq \sqrt[4]{2}$. Assume $x_k \geq \sqrt[4]{2}$ for some $k \geq 2$. Then

$$x_{k+1} = \frac{1}{2}\left(x_k + \frac{\sqrt{2}}{x_k}\right) \geq \sqrt[4]{2}$$

since $x_k > 0$ and $(x_k - \sqrt[4]{2})^2 \geq 0$. It follows that $(x_n)_{n \geq 2}$ is decreasing and bounded below by $\sqrt[4]{2}$, and hence convergent to its greatest lower bound. Since any convergent sequence is Cauchy (exercise!), we conclude that (x_n) is Cauchy. Let $L = \lim_{n \to \infty} x_n$. We have $L = \frac{1}{2}(L + \frac{\sqrt{2}}{L})$, hence $L = \sqrt[4]{2}$. $\qquad\square$

Lemma 11.55. *Let $a = (a_n)$ be a Cauchy sequence of rational numbers and let L be the real number which it defines. Then (a_n) can be viewed as a Cauchy sequence of real numbers and $\lim_{n \to \infty} a_n = L$.*

Proof. The first assertion is clear. Let $\varepsilon > 0$ and let $\delta \in \mathbb{Q}$ with $0 < \delta < \varepsilon$. Since (a_n) is Cauchy, there is N such that $|a_m - a_n| < \delta$ for all $m, n \geq N$. Consider $x_n = \widehat{a_n}$. Then $x_n - L \leq \delta$ for $n \geq N$. Similarly, $L - x_n \leq \delta$ for $n \geq N$. We conclude that for all $n \geq N$ we have $|x_n - L| \leq \delta < \varepsilon$, hence $\lim_{n \to \infty} a_n = \lim_{n \to \infty} x_n = L$. $\qquad\square$

Theorem 11.56. *Any Cauchy sequence of real numbers is convergent.*

Proof. Assume that $x = (x_n)$ is a Cauchy sequence of real numbers. For each positive integer n we have $x_n < x_n + 1/n$, hence we may find $a_n \in \mathbb{Q}$ with $x_n < a_n < x_n + 1/n$. Given $\varepsilon > 0$ let N_1 be an integer with $N_1 > \varepsilon/3$. For $n \geq N_1$ we have $|x_n - a_n| < \varepsilon/3$. The inequality

$$|a_m - a_n| \leq |a_m - x_m| + |x_m - x_n| + |x_n - a_n|$$

proves that $a = (a_n)$ is a Cauchy sequence, hence it defines a real number L such that $\lim_{n \to \infty} a_n = L$. For the fixed $\varepsilon > 0$ there is N_2 such that $|a_n - L| < \varepsilon/2$ for all $n \geq N_2$. It follows that for $n \geq \max\{N_1, N_2\}$ we get

$$|x_n - L| \leq |x_n - u_n| + |a_n - L| < \varepsilon/3 + \varepsilon/2 < \varepsilon,$$

hence $L = \lim_{n \to \infty} x_n$. $\qquad\square$

This property of real numbers is called *completeness*. We have seen that the set of rational numbers does not have this property, since there are Cauchy sequences of rational numbers converging to irrational numbers. We say that the set of real numbers was obtained by completing the set of rational numbers. The set of real numbers forms an Archimedean complete ordered field.

The set of rational numbers can be completed in a different way to obtain what is called the field of p-adic numbers, denoted $\mathbb{Q}_p$ (here $p \geq 2$ is a positive prime number). For the interested reader, we sketch here the construction of the p-adic numbers.

Fix a prime p. Each nonzero $x \in \mathbb{Q}$ can be written in a unique way as $x = p^n \cdot \dfrac{a}{b}$ where $n, a, b \in \mathbb{Z}, b > 0$ and a, b are not divisible by p. Define

$$|x|_p = p^{-n}.$$

By definition $|0|_p = 0$. For example, $\left|\dfrac{3}{4}\right|_2 = 2^2$ and $\left|\dfrac{8}{5}\right|_2 = 2^{-3}$. It can be proven that

$$|x + y|_p \leq \max\{|x|_p, |y|_p\} \leq |x|_p + |y|_p$$

and that

$$|xy|_p = |x|_p |y|_p$$

for all $x, y \in \mathbb{Q}$.

A sequence of rational numbers $x = (x_n)$ is Cauchy with respect to $|\cdot|_p$ if for any positive rational $\varepsilon > 0$ there is $n \in \mathbb{N}$ such that $|x_m - x_n|_p < \varepsilon$ for all $m, n \geq N$. The set $\mathbb{Q}_p$ is defined as the set of equivalence classes of Cauchy sequences of rational numbers with respect to $|\cdot|_p$.

11.3   Decimal representation of real numbers

Theorem 11.57. *(Decimal representation). Given a real number x there is a sequence $d_0, d_1, d_2, \ldots$ of integers uniquely determined by x such that*

(i) $d_0 = \lfloor x \rfloor$, the largest integer less than or equal to x;

(ii) $0 \leq d_n \leq 9$ for all $n \geq 1$, in fact $d_n = \lfloor 10^n x \rfloor - 10\lfloor 10^{n-1} x \rfloor$ for $n \geq 1$, where $\lfloor \cdot \rfloor$ denotes the floor function; and

(iii) the sequence defined inductively by

$$y_0 = d_0, \quad y_{n+1} = y_n + \frac{d_{n+1}}{10^{n+1}}$$

is Cauchy and $\lim\limits_{n \to \infty} y_n = x$. We write

$$x = d_0.d_1 d_2 \cdots d_n \cdots.$$

The terminating zeros are usually omitted.

Proof. It suffices to consider the case $x \geq 0$. Since $d_0 = \lfloor x \rfloor$, we have $10x = 10d_0 + x_1$, where $0 \leq x_1 < 10$. Let $d_1 = \lfloor x_1 \rfloor$, so $10x_1 = 10d_1 + x_2$ for some x_2 with $0 \leq x_2 < 10$. In general we can find x_n such that $10x_{n-1} = 10d_{n-1} + x_n$ and we set $d_n = \lfloor x_n \rfloor$. We can write

$$x = d_0 + \frac{d_1}{10} + \frac{d_2}{10^2} + \cdots + \frac{d_n}{10^n} + \frac{x_{n+1}}{10^{n+1}},$$

where $0 \leq x_{n+1} < 10$, hence

$$0 \leq x - \left(d_0 + \frac{d_1}{10} + \frac{d_2}{10^2} + \cdots + \frac{d_n}{10^n} \right) = x - y_n < \frac{1}{10^n}.$$

We conclude that $|x - y_n| < 10^{-n}$ and that $\lim_{n \to \infty} y_n = x$.

For the uniqueness part, note that of the two possible decimal representations of numbers of the form $\frac{a}{10^b}$ where a, b are nonnegative integers, the theorem chooses the one which consists of all zeros after a certain step. For example, $\frac{2}{10} = 0.2$ as opposed to $0.1999...$ □

Theorem 11.58. *A real number is rational if and only if its decimal representation terminates or has a group of digits that repeats infinitely many times.*

Proof. We have seen in Theorem 10.40 in the previous chapter that any rational number has a decimal representation which terminates or has periodicity.

Conversely, if the decimal expansion of $r \in \mathbb{R}$ terminates, say $r = 0.a_1 a_2 \cdots a_k$ where a_i are digits and $a_k \neq 0$, then

$$r = \frac{a_1 10^{k-1} + a_2 10^{k-2} + \cdots a_k}{10^k}$$

is a rational number. If r has a repeating group of decimals $b_1 b_2 \cdots b_m$ right after the decimal point, say $r = 0.\overline{b_1 b_2 \cdots b_m}$, then

$$10^m r = b_1 b_2 \cdots b_m + r, \quad r = \frac{b_1 b_2 \cdots b_m}{10^m - 1} \in \mathbb{Q}.$$

If $r = 0.a_1 a_2 \cdots a_k \overline{b_1 b_2 \cdots b_m}$, then

$$10^k r = a_1 a_2 \cdots a_k + 0.\overline{b_1 b_2 \cdots b_m},$$

$$r = \frac{\dfrac{b_1 b_2 \cdots b_m}{10^m - 1} + a_1 a_2 \cdots a_k}{10^k} = \frac{a_1 a_2 \cdots a_k b_1 b_2 \cdots b_m - a_1 a_2 \cdots a_k}{(10^m - 1)10^k} \in \mathbb{Q}.$$

□

Example 11.59.

$$0.\overline{1234} = \frac{1234}{9999}, \quad 0.123\overline{4567} = \frac{1234567 - 123}{9999000} = \frac{1234444}{9999000}.$$

11.4 Algebraic and transcendental numbers

Definition 11.60. *A real number α is* algebraic *if there is a polynomial $p(x) = a_n x^n + a_{n-1} x^{n-1} + \cdots a_1 x + a_0$ with integer coefficients such that $p(\alpha) = 0$. It can be proved that the set $\mathbb{A}$ of algebraic numbers is countable (in bijection with $\mathbb{N}$). A number which is not algebraic is called* transcendental. *The set of transcendental numbers is uncountable.*

Example 11.61. *Rational numbers are algebraic. Indeed, $\dfrac{a}{b}$ is a root of the polynomial $p(x) = bx - a$.*

Some irrational numbers are algebraic, for example $\sqrt{2}$ is a root of $p(x) = x^2 - 2$.

Also, $a = \sqrt{2} + \sqrt{3}$ is algebraic. To find a polynomial p with integer coefficients such that $p(a) = 0$, we need to compute

$$(x - (\sqrt{2} + \sqrt{3}))(x + (\sqrt{2} + \sqrt{3}))(x - (\sqrt{2} - \sqrt{3}))(x + (\sqrt{2} - \sqrt{3})) =$$

$$= (x^2 - (\sqrt{2} + \sqrt{3})^2)(x^2 - (\sqrt{2} - \sqrt{3})^2) = (x^2 - 5 - 2\sqrt{6})(x^2 - 5 + 2\sqrt{6}) =$$

$$= (x^2 - 5)^2 - (2\sqrt{6})^2 = x^4 - 10x^2 + 25 - 24 = x^4 - 10x^2 + 1,$$

so $p(x) = x^4 - 10x^2 + 1$.

The numbers e and π are transcendental. Proving that a certain number is transcendental is generally hard.

Example 11.62. *Euler's constant is defined as*

$$\gamma = \lim_{n \to \infty} (1 + \frac{1}{2} + \cdots + \frac{1}{n} - \ln n).$$

The existence of the limit can be proved using the mean value theorem. It is not known if γ is rational or irrational, algebraic or transcendent.

11.5 Complex numbers

Complex numbers were invented to solve polynomial equations like $x^2 + 1 = 0$. You may be familiar with the quadratic formula for solving $ax^2 + bx + c = 0$ for $a \neq 0$. If $b^2 - 4ac < 0$, you get something negative under the square root, and the solutions are complex numbers. We give a formal definition and prove some elementary properties of the set $\mathbb{C}$ of complex numbers. It turns out that $\mathbb{C}$ is an *algebraically closed field*: any polynomial equation with complex coefficients has all the roots in $\mathbb{C}$.

Definition 11.63. *The set of complex numbers is*

$$\mathbb{C} = \mathbb{R} \times \mathbb{R} = \{\langle a, b \rangle : a, b \in \mathbb{R}\}$$

with operations

$$\langle a, b \rangle + \langle c, d \rangle = \langle a + c, b + d \rangle \text{ and } \langle a, b \rangle \cdot \langle c, d \rangle = \langle ac - bd, ad + bc \rangle.$$

Theorem 11.64. *Addition $+$ and multiplication $\cdot$ are commutative and associative. Multiplication is distributive with respect to addition. For $z \in \mathbb{C}$, we have*

$$z + \langle 0, 0 \rangle = z, \quad z \cdot \langle 0, 0 \rangle = \langle 0, 0 \rangle, \quad z \cdot \langle 1, 0 \rangle = z.$$

For each $z \in \mathbb{C}$ there is a unique w such that $z + w = \langle 0, 0 \rangle$. We have cancellation $w + z = u + z \Rightarrow w = u$.

Proof. Commutativity and associativity of addition are easy verifications. We have

$$\langle a, b \rangle \cdot \langle c, d \rangle = \langle ac - bd, ad + bc \rangle = \langle ca - db, da + cb \rangle = \langle c, d \rangle \cdot \langle a, b \rangle.$$

$$(\langle a, b \rangle \cdot \langle c, d \rangle) \cdot \langle e, f \rangle = \langle ac - bd, ad + bc \rangle \cdot \langle e, f \rangle =$$

$$= \langle ace - bde - adf - bcf, acf - bdf + ade + bce \rangle =$$

$$\langle a, b \rangle \cdot \langle ce - df, cf + de \rangle = \langle a, b \rangle \cdot (\langle c, d \rangle \cdot \langle e, f \rangle).$$

$$\langle a, b \rangle \cdot (\langle c, d \rangle + \langle e, f \rangle) = \langle a, b \rangle \cdot \langle c + e, d + f \rangle = \langle ac + ae - bd - bf, ad + af + bc + be \rangle =$$

$$\langle ac - bd, ad + bc \rangle + \langle ae - bf, af + be \rangle = \langle a, b \rangle \cdot \langle c, d \rangle + \langle a, b \rangle \cdot \langle e, f \rangle.$$

We leave the remaining properties as an exercise. □

Definition 11.65. *If $z = \langle a, b \rangle \in \mathbb{C}$, define the conjugate $\bar{z} = \langle a, -b \rangle$ and the absolute value of z by $|z| = \sqrt{a^2 + b^2}$. Note that $|z| \in [0, \infty)$.*

A complex number $\langle a, b \rangle$ can be identified with a point in the Euclidean plane $\mathbb{R}^2$ which has coordinates a and b. Since $\mathbb{C}$ is identified with $\mathbb{R}^2$, this plane is also called the complex plane.

The operations with complex numbers have geometric interpretation. For example, the sum of two complex numbers is obtained by the parallelogram rule used for addition of vectors. The complex conjugate can be obtained by reflecting in the x-axis. The absolute value $|z - w|$ represents the distance between z and w as points in the plane. Multiplication with $\langle 0, 1 \rangle$ corresponds to a rotation of $90°$, etc.

There are several geometric properties that can be interpreted using complex numbers.

Example 11.66. *The equation $|z + w|^2 + |z - w|^2 = 2(|z|^2 + |w|^2)$ says that the sum of the squares of the diagonals of a parallelogram equals the sum of the squares of the sides.*

Theorem 11.67. *We have*
1. $|z| = 0$ *iff* $z = \langle 0, 0 \rangle$.
2. $|z \cdot w| = |z| \cdot |w|$.
3. $z\bar{z} = |z|^2$ *and* $|z| = |\bar{z}|$.
4. $z \cdot w = \langle 0, 0 \rangle$ *iff* $z = \langle 0, 0 \rangle$ *or* $w = \langle 0, 0 \rangle$.
5. $\overline{z + w} = \bar{z} + \bar{w}$.
6. $\overline{zw} = \bar{z}\bar{w}$.
7. $|a| = |\langle a, 0 \rangle| \le |\langle a, b \rangle|$.
8. $|z + w| \le |z| + |w|$.

Proof. The first six properties are easy to check using the definition.
7. We have $|a| = \sqrt{a^2} \le \sqrt{a^2 + b^2}$.
8. Let $z = \langle a, b \rangle$ and $w = \langle c, d \rangle$. We have

$$|z + w|^2 = (z + w)(\bar{z} + \bar{w}) = z\bar{z} + z\bar{w} + \bar{z}w + w\bar{w} =$$

$$= |z|^2 + \langle ac + bd, bc - ad \rangle + \langle ac + bd, ad - bc \rangle + |w|^2 =$$

$$|z|^2 + 2\langle ac + bd, 0 \rangle + |w|^2 \le |z|^2 + 2|z\bar{w}| + |w|^2 =$$

$$= |z|^2 + 2|z||w| + |w|^2 = (|z| + |w|)^2.$$

The inequality follows by taking square roots. $\qquad\square$

Theorem 11.68. *For each $z \ne \langle 0, 0 \rangle$ there exists a unique w such that $z \cdot w = \langle 1, 0 \rangle$.*
If $z \ne \langle 0, 0 \rangle$ and $z \cdot w = z \cdot u$, then $w = u$.

Proof. Let $z = \langle a, b \rangle$ with $a^2 + b^2 \ne 0$. Consider $w = \langle \dfrac{a}{a^2 + b^2}, -\dfrac{b}{a^2 + b^2} \rangle$.
Then $z \cdot w = \langle 1, 0 \rangle$. Note that $w = \dfrac{\bar{z}}{|z|^2}$. This w is called the inverse of z,
denoted z^{-1}. For the second property, multiply both sides by z^{-1}. $\qquad\square$

Since $\langle a, 0 \rangle + \langle b, 0 \rangle = \langle a + b, 0 \rangle$ and $\langle a, 0 \rangle \cdot \langle b, 0 \rangle = \langle ab, 0 \rangle$, we obtain the following corollary.

Corollary 11.69. *The field $(\mathbb{R}, +, \cdot)$ is isomorphic with the subset $\{\langle a, 0 \rangle : a \in \mathbb{R}\}$ of $\mathbb{C}$ under addition and multiplication. We identify a real number a with the pair $\langle a, 0 \rangle \in \mathbb{C}$.*

Denote by i the element $\langle 0, 1 \rangle$ of $\mathbb{C}$. Then we have

$$i^2 = \langle 0, 1 \rangle \cdot \langle 0, 1 \rangle = \langle -1, 0 \rangle = -1.$$

Any complex number $\langle a, b \rangle$ can be written as $\langle a, b \rangle = \langle a, 0 \rangle + \langle b, 0 \rangle \cdot \langle 0, 1 \rangle = a + bi$. Note that $\overline{a + bi} = a - bi$. The real numbers a and b are called the *real part* and the *imaginary part* of $z = a + bi$. We write $a = \text{Re}(z)$, $b = \text{Im}(z)$.

Theorem 11.70. *There is no order relation on $\mathbb{C}$ that extends the usual order relation on $\mathbb{R}$, hence $\mathbb{C}$ it not an ordered field.*

Proof. Indeed, if such an order $\preceq$ exists, then $z^2 \succeq 0$ for all $z \in \mathbb{C}$. But $i^2 = -1 \prec 0$, a contradiction. $\qquad\square$

The quadratic equation $ax^2 + bx + c = 0$ with complex coefficients can be solved using the quadratic formula,

$$x_{1,2} = \frac{-b + \sqrt{b^2 - 4ac}}{2a},$$

except this involves the square root of a complex number, which in general has two possible complex values.

Example 11.71. *Let's find $\sqrt{2 + i} = a + bi$ with $a, b \in \mathbb{R}$. We have $(a + bi)^2 = 2 + i$, so $a^2 - b^2 + 2abi = 2 + i$. The system*

$$a^2 - b^2 = 2$$

$$2ab = 1$$

has two real solutions

$$a = \sqrt{\frac{2 + \sqrt{5}}{2}}, \quad b = \frac{1}{\sqrt{4 + 2\sqrt{5}}}$$

and

$$a = -\sqrt{\frac{2 + \sqrt{5}}{2}}, \quad b = -\frac{1}{\sqrt{4 + 2\sqrt{5}}}.$$

11.6 The trigonometric form of a complex number

For various computations, including finding the n-th roots of a complex number, the trigonometric form is very useful. It is similar to polar coordinates in the plane.

Theorem 11.72. *(Trigonometric form) If $z \in \mathbb{C} \setminus \{0\}$, then there is $r > 0$ and $w \in \mathbb{C}$ with $|w| = 1$ such that $z = rw$. The number w is of the form $\cos t + i \sin t$ with $t \in [0, 2\pi)$, so*

$$z = |z|(\cos t + i \sin t).$$

Proof. Consider $r = |z|$ and $w = \dfrac{z}{|z|}$. Then $|w| = 1$ and w is on the trigonometric circle $x^2 + y^2 = 1$, hence there is $t \in [0, 2\pi)$ such that $w = \cos t + i \sin t$. The number $r \geq 0$ is the distance from z to the origin, sometimes called the polar radius, and the angle t is the polar angle, also called the *argument* of z. $\qquad\square$

Example 11.73. *Let's find the trigonometric form of $z = -2 + 2i$. We have*

$$|z| = \sqrt{4+4} = 2\sqrt{2}, \quad \frac{z}{|z|} = -\frac{1}{\sqrt{2}} + \frac{1}{\sqrt{2}}i,$$

so $t = \dfrac{3\pi}{4}$ *and*

$$-2 + 2i = 2\sqrt{2}\left(\cos\frac{3\pi}{4} + i\sin\frac{3\pi}{4}\right).$$

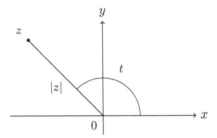

Corollary 11.74. *Two complex numbers in trigonometric form* $z_n = r_n(\cos t_n + i\sin t_n), n = 1, 2$ *can be multiplied by the formula*

$$z_1 z_2 = r_1 r_2(\cos(t_1 + t_2) + i\sin(t_1 + t_2)).$$

Proof. Indeed,

$$(\cos t_1 + i\sin t_1)(\cos t_2 + i\sin t_2) =$$
$$= (\cos t_1 \cos t_2 - \sin t_1 \sin t_2) + i(\sin t_1 \cos t_2 + \sin t_2 \cos t_1) =$$
$$= \cos(t_1 + t_2) + i\sin(t_1 + t_2).$$

$\square$

Theorem 11.75. *We have the formula (De Moivre)*

$$(\cos t + i\sin t)^n = \cos nt + i\sin nt.$$

Proof. By induction. Clearly this is true for $n = 1$. Assume

$$(\cos t + i\sin t)^k = \cos kt + i\sin kt$$

for some $k \geq 1$. Then

$$(\cos t + i\sin t)^{k+1} = (\cos t + i\sin t)^k(\cos t + i\sin t) = (\cos kt + i\sin kt)(\cos t + i\sin t) =$$
$$= (\cos kt \cos t - \sin kt \sin t) + i(\cos kt \sin t + \sin kt \cos t) = \cos(k+1)t + i\sin(k+1)t.$$

$\square$

Corollary 11.76. *If* $w = |w|(\cos t + i \sin t)$ *with* $t \in [0, 2\pi)$, *then the equation* $z^n = w$ *for* n *a positive integer has* n *complex roots*

$$z_k = \sqrt[k]{|w|}\left(\cos\frac{2k\pi + t}{n} + i\sin\frac{2k\pi + t}{n}\right), k = 0, 1, ..., n - 1.$$

Example 11.77. *Let's find all cubic roots of* i. *Since* $i = \cos\dfrac{\pi}{2} + i\sin\dfrac{\pi}{2}$, *we get*

$$z_1 = \cos\frac{\pi}{6} + i\sin\frac{\pi}{6} = \frac{\sqrt{2}}{2} + \frac{1}{2}i,$$

$$z_2 = \cos\frac{5\pi}{6} + i\sin\frac{5\pi}{6} = -\frac{\sqrt{2}}{2} + \frac{1}{2}i,$$

$$z_3 = \cos\frac{3\pi}{2} + i\sin\frac{3\pi}{2} = -i.$$

Note that z_1, z_2, z_3 *are the vertices of an equilateral triangle.*

Theorem 11.78. *The field* $\mathbb{C}$ *is complete (every Cauchy sequence is convergent).*

Proof. Let (z_n) with $z_n = a_n + b_n i$ be a Cauchy sequence in $\mathbb{C}$. This means that given $\varepsilon > 0$ arbitrary, there is $N \in \mathbb{N}$ such that $|z_m - z_n| < \varepsilon$ for all $m, n \geq N$. Since $|a_m - a_n|, |b_m - b_n| \leq |z_m - z_n|$, we conclude that the sequences (a_n) and (b_n) are Cauchy. Since $\mathbb{R}$ is complete, we get $a_n \to a$ and $b_n \to b$. Then $z_n \to a + bi$. □

It is possible to extend the complex number field to a larger system of numbers denoted by $\mathbb{H}$. In fact, one can define addition and multiplication on $\mathbb{H} = \mathbb{C} \times \mathbb{C}$ to get a number system called the *quaternions*, invented by Hamilton. More precisely,

$$\langle z_1, w_1 \rangle + \langle z_2, w_2 \rangle = \langle z_1 + z_2, w_1 + w_2 \rangle,$$

$$\langle z_1, w_1 \rangle \cdot \langle z_2, w_2 \rangle = \langle z_1 z_2 - w_1 \bar{w}_2, z_1 w_2 + \bar{z}_2 w_1 \rangle.$$

We can identify $z \in \mathbb{C}$ with $\langle z, 0 \rangle \in \mathbb{H}$. It can be proved that the multiplication of quaternions is associative but not commutative. Any nonzero quaternion has a multiplicative inverse. The set $\mathbb{H}$ becomes a structure called a *division ring* or *skew field*.

It is possible to give $\mathbb{H} \times \mathbb{H}$ additive and multiplicative operations, obtaining the *octonions* or *Cayley numbers*. Multiplication is both noncommutative and nonassociative. The quaternions and the Cayley numbers are used in mathematical physics, group representation, and algebraic topology.

11.7 Exercises

Exercise 11.79. *Show that $C = \{x \in \mathbb{Q} : x \geq 1\}$ is not a cut.*

Exercise 11.80. *Prove the following:*
    a. If C is a cut, then C is positive if and only if $0 \in C$.
    b. If C is a positive cut and D is a negative cut, then $D \prec C$.
    c. If C is a negative cut and $D \prec C$, then D is a negative cut.
    d. $C \prec \hat{0}$ if and only if C is a negative cut.

Exercise 11.81. *Prove that for $r, s \in \mathbb{Q}$ we have $\hat{r} \cdot \hat{s} = \widehat{rs}$, where $\hat{r}$ is the cut $\{x \in \mathbb{Q} : x < r\}$.*

Exercise 11.82. *Prove that for the cut D in Example 11.5 we have $D \cdot D = \hat{2}$.*

Exercise 11.83. *Let $C = \{x \in \mathbb{Q} : x^3 < 5\}$. Prove that C is a cut. (Hint: Given $s \in C$, show that $t = \dfrac{15s}{s^3 + 10}$ belongs to C and that $s < t$.)*

Exercise 11.84. *Let C be a cut and let $s \in \mathbb{Q}$ such that $s > t$ for all $t \in C$. Prove that $s \notin C$.*

Exercise 11.85. *Prove that there is no rational number x such that $x^3 = 7$. Construct a real number x with this property.*

Exercise 11.86. *Let $S \subseteq \mathbb{R}$ be a set which is bounded below. Use quantifiers to describe the fact that $b = \inf S$.*

Exercise 11.87. *Prove that between any two real numbers there is a rational and an irrational number.*

Exercise 11.88. *Prove that $\sqrt{2} + \sqrt{3} - \sqrt{5 + 2\sqrt{6}}$ is rational.*

Exercise 11.89. *Prove that*
    1. A sequence of real numbers (x_n) has at most one limit.
    2. A convergent sequence (x_n) of real numbers is Cauchy.

Exercise 11.90. *Prove that the sequences x, y with*

$$x_n = 1 - \frac{1}{3} + \frac{1}{5} - \cdots + \frac{(-1)^n}{2n+1}, \quad y_n = 1 + \frac{1}{1!} + \frac{1}{2!} + \cdots + \frac{1}{n!}$$

are Cauchy sequences of rational numbers.

Exercise 11.91. *Prove that a decreasing sequence of real numbers which is bounded below is convergent. Deduce that an increasing sequence of real numbers which is bounded above is convergent.*

Exercise 11.92. *Show that $1 - \sqrt{3}, \sqrt{5} - \sqrt{2}$ are algebraic.*

Exercise 11.93. *Prove that if α is algebraic, then $k\alpha$ is algebraic for any integer $k \neq 0$. Also, if $\alpha \neq 0$ is algebraic, then the inverse $1/\alpha$ is algebraic.*

Exercise 11.94. *Prove that if β is transcendental, then $1/\beta$ is also transcendental.*

Exercise 11.95. *Determine i^n for $n \in \mathbb{N}$ where $i^2 = -1$.*

Exercise 11.96. *Prove that*
 a. $z + \bar{z} = 2\,Re(z)$ *and* $z - \bar{z} = 2i\,Im(z)$.
 b. $|Re(z)| \leq |z|$ *and* $|Im(z)| \leq |z|$.

Exercise 11.97. *Let $z = 3 + 4i$ and $w = 5 - 2i$. Evaluate and simplify*

$$|z|, |w|, z + w, zw, z/w, \overline{iz - 3w}.$$

Determine the real part and the imaginary part of each number.

Exercise 11.98. *A complex number α is called algebraic if it satisfies a polynomial equation with integer coefficients*

$$a_n x^n + a_{n-1} x^{n-1} + \cdots + a_1 x + a_0 = 0.$$

For example, i is algebraic since it satisfies $x^2 + 1 = 0$. Assuming that each polynomial equation of degree n has n complex roots, prove that the set of algebraic complex numbers is countable.

Exercise 11.99. *If z_1, z_2, z_3 are points in the complex plane which determine a triangle, show that the medians of this triangle intersect at the point $\frac{1}{3}(z_1 + z_2 + z_3)$.*

Exercise 11.100. *Given that $z, w \in \mathbb{C}$ are two vertices of a square, find the other vertices in all possible cases.*

Exercise 11.101. *Solve the quadratic equations*

$$a)\ x^2 - ix + 2 = 0,\quad b)\ ix^2 + (1 - i)x + 1 = 0.$$

Exercise 11.102. *Let $z = a + bi, w = c + di \in \mathbb{C}$. Define $z \prec w$ if $a < c$ or $a = c$ and $b < d$ (lexicographic strict order). Does the order $\preceq$ on $\mathbb{C}$ have the least upper bound property?*

Exercise 11.103. *Show that $\cos t + i \sin t + 1 = 2 \cos \dfrac{t}{2} \left(\cos \dfrac{t}{2} + i \sin \dfrac{t}{2} \right)$.*

Exercise 11.104. *If $z \in \mathbb{C}$ such that $|z| = 1$, compute $|1 + z|^2 + |1 - z|^2$.*

Exercise 11.105. *Write $w = \sqrt{2} - i\sqrt{2}$ in trigonometric form and solve the equation $z^4 = w$.*

Exercise 11.106. *Prove that if z_1, z_2, z_3 are the vertices of an equilateral triangle, then $z_1^2 + z_2^2 + z_3^2 = z_1 z_2 + z_1 z_3 + z_2 z_3$.*

Exercise 11.107. *Let $n \geq 2$. If $z_0, z_1, ..., z_{n-1}$ are the roots of $z^n = 1$, prove that $z_0 + z_1 + \cdots + z_{n-1} = 0$.*

Exercise 11.108. *Suppose $z \in \mathbb{C}$ such that $z + \dfrac{1}{z} = 1$. Find $z^2 + \dfrac{1}{z^2}$ and in general $z^n + \dfrac{1}{z^n}$ for $n \geq 2$.*

Exercise 11.109. *Compute $(1 - i)^{-20}$ and $(1 + i)^{75}$.*

Exercise 11.110. *Compute $(\cos t + i \sin t)^3$ in two different ways to show that*

$$\cos 3t = \cos^3 t - 3 \cos t \sin^2 t, \quad \sin 3t = 3 \cos^2 t \sin t - \sin^3 t.$$

Exercise 11.111. *Let $z \in \mathbb{C} \setminus \{0\}$. Prove that $\dfrac{z}{|z|} + \dfrac{|z|}{z}$ is a real number.*

Exercise 11.112. *Denote $j = \langle 0, 1 \rangle$ and $k = \langle 0, i \rangle$ as elements in $\mathbb{H} = \mathbb{C} \times \mathbb{C}$. Prove that*

$$j \cdot j = k \cdot k = -1, i \cdot j = k, j \cdot k = i, k \cdot i = j, j \cdot i = -k, k \cdot j = -i, i \cdot k = -j.$$

Exercise 11.113. *Show that any quaternion $\langle z, w \rangle \in \mathbb{H}$ can be written in the form $z + wj$.*

Bibliography

[1] Beck, M. and Geoghegan, R., *The Art of Proof*, Springer, 2011.

[2] Birkhoff, G. and Mac Lane, S., *Algebra*, AMS Chelsea Publishing, 1999.

[3] Ciesielski, K., *Set Theory for the Working Mathematician*, Cambridge University Press, 1997.

[4] Davidson, K.R. and Donsig, A.P. *Real Analysis and Applications. Theory in Practice*, Undergraduate Texts in Mathematics. Springer, New York, 2010.

[5] Daepp, U. and Gorkin, P. *Reading, Writing and Proving. A Closer Look at Mathematics*, Springer, 2003.

[6] Engel, A. *Problem-Solving Strategies*, Springer, 1998.

[7] Folland, G. *Real Analysis. Modern Techniques and Their Applications, Second Edition*, John Wiley & Sons, Inc. 1999.

[8] Galovich, S. *Doing Mathematics. An Introduction to Proofs and Problem Solving, Second Edition*, Thomson Brooks/Cole, 2007.

[9] Halmos, P.R., *Naive Set Theory*, Springer-Verlag, 1974.

[10] Hammack, R. *Book of Proof*, Richard Hammack, 2009.

[11] Krantz, S. *Real Analysis and Foundations*, CRC Press, 2000.

[12] Krantz, S. *The Elements of Advanced Mathematics, 2nd Edition*, Chapman & Hall, 2002.

[13] Liebeck, M. *A Concise Introduction to Pure Mathematics*, CRC Press, 2011.

[14] Pedersen, G.K. *Analysis Now*, Springer, 1989.

[15] Rudin, W. *Principles of Mathematical Analysis*, McGraw-Hill, 1976.

[16] Stoll, R. *Set Theory and Logic*, Dover, 1979.

[17] Sundstrom, T. *Mathematical Reasoning: Writing and Proof*, CreateSpace Independent Publishing Platform, 2013.

[18] Youse, B.K., *The Number System*, Dickenson Publishing Company, 1965.

Answers to select exercises

Ex. 1.24. For example, we can form the following statements:
1) If $2 < 3$ then $((1 + 1 = 2)$ or $(3 + 2 = 6$ and $5 > 7))$.
2) If $2 < 3$ then $((1 + 1 = 2$ or $3 + 2 = 6)$ and $(5 > 7))$.
3) (If $2 < 3$ then $1 + 1 = 2)$ or $(3 + 2 = 6$ and $5 > 7)$.
4) ((If $2 < 3$ then $1 + 1 = 2)$ or $3 + 2 = 6)$ and $5 > 7$.
The statements 1 and 3 are true, but statements 2 and 4 are false.

Ex. 1.25. The conditionals "If $0 < 0$ then $0 > 0$" and "If $1 < 0$ then $1 > 0$" are true because a false statement implies anything.

Ex. 1.28. a) yes; b) no; c) no; d) no.

Ex. 1.29. Contrapositive: "If $n^2 - n - 6$ is odd, then n is even". Converse: "If $n^2 - n - 6$ is even, then n is odd".

Ex. 1.30. It is logically equivalent with $\neg Q$.

Ex. 1.31. a) not valid; b) valid; c) not valid.

Ex. 1.32. I will not be admitted to the university.

Ex. 1.33. 1) The converse is false since $\lim_{n \to \infty} \dfrac{1}{n} = 0$ but $\sum_{n=1}^{\infty} \dfrac{1}{n} = \infty$;

2) The converse is true because of the mean value theorem;

3) The converse is false since $\sum_{n=1}^{\infty} \dfrac{(-1)^n}{n}$ is convergent by the alternating series test, but $\sum_{n=1}^{\infty} \dfrac{1}{n} = \infty$.

4) The converse is false since for example $f : [a, b] \to \mathbb{R}, f(x) = 1$ for $a < x \le b$ and $f(a) = 0$ attains the maximum of 1, but it is not continuous.

Ex. 1.34. a) $\forall n \in \mathbb{Z}$ we have n is odd (false); $\exists n \in \mathbb{Z}$ such that n is even (true).

b) $\exists r \in \mathbb{Q}$ such that $r^2 = 2$ (false); $\forall r \in \mathbb{Q}$ we have $r^2 \ne 2$ (true).

c) $\forall x \in \mathbb{R} \; \exists! \; y \in \mathbb{R}$ such that $y^3 = x$ (true); $\exists x \in \mathbb{R} \; \forall y \in \mathbb{R}$ we have $y^3 \ne x$ or $\exists x \in \mathbb{R} \; \exists y_1, y_2 \in \mathbb{R}$ with $y_1 \ne y_2$ and $y_1^3 = y_2^3 = x$ (false).

d) $\forall n \in \mathbb{Z}$ odd $\exists m \in \mathbb{Z}$ even such that m divides n (false); $\exists n \in \mathbb{Z}$ odd $\forall m \in \mathbb{Z}$ even we have m does not divide n (true).

e) $\forall x \in \mathbb{Q}$ we have $x^2 + 1 \ne 0$ (true); $\exists x \in \mathbb{Q}$ such that $x^2 + 1 = 0$ (false).

f) $\exists x \; \exists y$ with $x = 3y$ (true); $\forall x \forall y$ we have $x \ne 3y$ (false).

g) If $\forall x, y \in \mathbb{R}$ and $\forall \varepsilon > 0$ we have $x < y + \varepsilon$, then $x \le y$ (true); if $\forall x, y \in \mathbb{R}$ and $\forall \varepsilon >$ we have $x < y + \varepsilon$ but $x > y$ (false).

h) $\forall x \notin \mathbb{Q}$ we have $\sqrt{x} \notin \mathbb{Q}$ (true); $\exists x \notin \mathbb{Q}$ such that $\sqrt{x} \in \mathbb{Q}$ (false).

Ex. 2.10. a) We prove: if $x \geq 0$, then $x^2 + 4x \geq 0$. Indeed, $x^2 + 4x = x(x+4) \geq 0$ since both factors x and $x + 4$ are positive. If we try a direct proof, we start from $x(x + 4) < 0$ and we have to analyze two cases: ($x < 0$ and $x + 4 > 0$) or ($x > 0$ and $x + 4 < 0$). Since the second case cannot occur, we conclude that $-4 < x < 0$, in particular $x < 0$.

b) We prove: if a, b have the same parity, then ab is even or $a + b$ is even. Indeed, if both a, b are even, then ab and $a + b$ are even. If both a, b are odd, then ab is odd but $a + b$ is even. A direct proof looks harder.

c) We prove: if a is even, then a^2 is divisible by 4. Indeed, let $a = 2k$. Then $a^2 = 4k^2$ is a multiple of 4. For a direct proof, we would have to analyze three cases $a^2 = 4k + 1$, $a^2 = 4k + 2$, or $a^2 = 4k + 3$ and conclude that a must be odd.

Ex. 2.15. a) Suppose we found some integers a and b such that $a^2 - 4b = 3$, hence $a^2 = 4b + 3$. But any integer a is either even or odd. In the first case, $a = 2k$ implies $a^2 = 4k$, a multiple of 4. In the second case, $a = 2k + 1$ implies $a^2 = 4k^2 + 4k + 1$, a multiple of 4 plus 1. We get a contradiction with the fact that a^2 is a multiple of 4 plus 3.

b) Assume $\sqrt{6} = a/b$ with $a, b \in \mathbb{Z}$ and $b \neq 0$. We may assume a, b relatively prime. We get $a^2 = 6b$, in particular a is even, say $a = 2n$. Then $4n^2 = 6b$, hence $2n^2 = 3b$ and b must also be even, a contradiction.

c) Assume $\sqrt{2} + \sqrt{3} = r \in \mathbb{Q}$. Squaring both sides, $5 + 2\sqrt{6} = r^2$ which gives $\sqrt{6} = \dfrac{r^2 - 5}{2} \in \mathbb{Q}$, a contradiction.

d) Assume a, b rational. Then ab is rational, a contradiction.

e) Assume a, b and $\sqrt{a} + \sqrt{b} = r$ rational, but say $\sqrt{a}$ irrational. Then from $\sqrt{b} = r - \sqrt{a}$, by squaring both sides, we get $b = r^2 + a - 2r\sqrt{a}$, or $\sqrt{a} = \dfrac{r^2 + a - b}{2r}$ which is rational, contradiction. In fact a direct proof is more appropriate here.

Ex. 2.27 a). For example $\sqrt{2} + (-\sqrt{2}) = 0$.

b) We have $\sqrt{2} \cdot \sqrt{2} = 2$.

c) Take $a = 3, b = 12$.

d) Let $a = b = 1$.

Ex. 2.36. The gap is in the step from one horse to two horses: horse 1 may have a different color than horse 2.

Ex. 2.40. a) Assuming the hypothesis, it follows that no matter what x is, one of the two statements $P(x)$ or $Q(x)$ is true. The converse is false, since $P(x)$ may be true exactly when $Q(x)$ is false. For example, let x be an integer, let $P(x)$ be the statement "x is even", and let $Q(x)$ be "x is odd". Then any integer is even or odd, hence $\forall x \in \mathbb{Z}(P(x) \vee Q(x))$ is true, but both $\forall x \in \mathbb{Z}\, P(x)$ and $\forall x \in \mathbb{Z}\, Q(x)$ are false, so $(\forall x \in \mathbb{Z}P(x)) \vee (\forall x \in \mathbb{Z}Q(x))$ is false.

b) Assuming that we found an x for which both $P(x)$ and $Q(x)$ are true, in particular we found an x for which $P(x)$ is true and the same x for which $Q(x)$ is true. The converse is false since we can take x to be a real number,

$P(x)$ to be $x > 0$ and $Q(x)$ to be $x < 0$. Then certainly there are positive numbers and there are negative numbers, but no real number is positive and negative in the same time.

Ex. 2.41. Let $ABCD$ be a quadrilateral and let K, L, M, N be the midpoints of the sides AB, BC, CD and DA, respectively. Then $KL \parallel AC, MN \parallel AC$ and $KL = MN = \frac{1}{2}AC$, so $KLMN$ is a parallelogram.

Ex. 2.43. Assuming $\sqrt{2} + \sqrt{6} \geq \sqrt{15}$ and squaring both sides we get $4\sqrt{3} \geq 15$ which is false, a contradiction.

Ex. 2.44. Assuming $\sqrt{3} = a\sqrt{2} + b$ for some rational numbers a and b we get $3 = 2a^2 + b^2 + 2ab\sqrt{2}$, which implies that $\sqrt{2}$ is rational, a contradiction.

Ex. 2.48. $x \in (-\infty, \frac{1-\sqrt{13}}{2}) \cup (\frac{1+\sqrt{13}}{2}, \infty)$.

Ex. 2.50. Take $n = 10$.

Ex. 2.51. Take $n = 11$.

Ex. 2.52. For $n = 2$ we take $x = y$ and get

$$f(2x) = \frac{f(x)^2}{2f(x)} = \frac{f(x)}{2}$$

since $f(x) > 0$, so the formula is true. Assume $f(kx) = f(x)/k$ for some $k \geq 2$. Then

$$f((k+1)x) = f(kx + x) = \frac{f(kx)f(x)}{f(kx) + f(x)} = \frac{f(x)^2}{k(f(x)/k + f(x))} = \frac{f(x)}{k+1}.$$

Ex. 2.64. The formula works for $n = 2$, since one line determines 2 regions. Assume k lines determine $(k^2 + k + 2)/2$ regions and consider one more line. Since the new line crosses all previous lines, it determines $k + 1$ more regions, for a total of

$$\frac{k^2 + k + 2}{2} + k + 1 = \frac{k^2 + 2k + 1 + k + 1 + 2}{2} = \frac{(k+1)^2 + (k+1) + 2}{2}.$$

Ex. 2.65. By induction over the number of lines. For one line, we color one region red and one blue, so the statement is true. Suppose that we have k lines which determine regions colored with red and blue as required. We want to show that one more line will determine regions that can be colored as required. If the new line does not contain an intersection point of the first k lines, it will split each intersecting region in two. If the intersecting region was red, we color the new region in blue and vice versa. If the new line goes through an intersection point of the first k lines, then it splits some old regions in two, and some old regions remain unchanged. For the splitted regions, we use the above method of coloring. For the other regions we may need to change some of the colors to obtain a coloring satisfying the requirement.

Ex. 3.45. $A = \{-2, -1, 0, 1, 2, 3, 4, 5, 6, 7, 8\}$.

Ex. 3.46. We have $|x - 2| = |x - 1 - 1| \leq |x - 1| + 1 < 1 + 1 = 2$, so $A \subseteq B$.

Ex. 3.47. We have $C = (2, 4)$ and $D = (4, 8)$, hence $C \cap D = \emptyset$.

Ex. 3.49. 1. We have $A \subseteq A \cup (A \cap B)$ and $A \cup (A \cap B) \subseteq A$ since each member of the union is a subset of A. By double inclusion, we get equality.

2. Let $x \in A \cup (B \cap C)$. This means $x \in A$ or ($x \in B$ and $x \in C$). By distributivity, this is equivalent to ($x \in A$ or $x \in B$) and ($x \in A$ or $x \in C$), and hence equivalent to $x \in (A \cup B) \cap (A \cup C)$.

Ex. 3.52. From the first condition, A and B are subsets of C. From the second condition, we get $C \subseteq B$, hence $B = C$. The last condition gives $B \subseteq A$, hence $C \subseteq A$ and therefore $A = C$.

Ex. 3.55. Consider $B_1 = A_1, B_2 = A_2 \setminus A_1, B_3 = A_3 \setminus (A_1 \cup A_2), ..., B_n = A_n \setminus (A_1 \cup A_2 \cup \cdots \cup A_{n-1})$. By construction, the sets B_i are disjoint and

$$\bigcup_{k=1}^{n} A_k = \bigcup_{k=1}^{n} B_k.$$

Ex. 3.56. The first condition implies that X and Y are subsets of $\{1, 2, 3, 4, 5, 6\}$. The second condition tells us that $1, 2, 3, 4 \in X$ and $1, 2, 3, 4 \in Y$. Since $\{4, 6\}$ is not a subset of X, $6 \notin X$, so it must be that $6 \in Y$. Condition d gives $5 \in X$, hence $X = \{1, 2, 3, 4, 5\}$ and $Y = \{1, 2, 3, 4, 6\}$.

Ex. 3.57. Since $X \triangle Y = (X \cup Y) \setminus (X \cap Y)$, we must have $X = \{5, 6\} \cup A$, and $Y = \{5, 6\} \cup (\{1, 2, 3, 4\} \setminus A)$ where $A \subseteq \{1, 2, 3, 4\}$. There are $2^4 = 16$ solutions.

Ex. 3.58. a. $X = \{1, 2\}$; b. $X = \{3\}$; c. $X = \{1, 2, 4\}$.

Ex. 4.71. The largest domain is $\mathbb{R} \setminus \{-3, 3\}$. Its graph has vertical asymptotes at $x = \pm 3$ and a horizontal asymptote at $y = 1$. Using the graph, we conclude that the range of f is $(-\infty, -1] \cup (1, \infty)$.

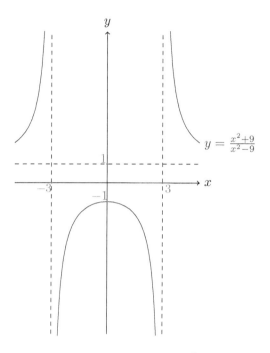

$$y = \frac{x^2+9}{x^2-9}$$

Ex. 4.72. We have $f + g : \mathbb{R} \to \mathbb{R}, (f + g)(x) = \begin{cases} 2x, & x \le 0 \\ 2x + 3, & 0 < x \le 3 \\ 4, & x > 3, \end{cases}$

$$f - g : \mathbb{R} \to \mathbb{R}, (f - g)(x) = \begin{cases} 4, & x \le 0 \\ 1, & 0 < x \le 3 \\ -2x + 2, & x > 3, \end{cases}$$

$$f \cdot g : \mathbb{R} \to \mathbb{R}, (f \cdot g)(x) = \begin{cases} x^2 - 4, & x \le 0 \\ x^2 + 3x + 2, & 0 < x \le 3 \\ -x^2 + 2x + 3, & x > 3, \end{cases}$$

$$f/g : \mathbb{R} \to \mathbb{R}, (f/g)(x) = \begin{cases} \dfrac{x+2}{x-2}, & x \le 0 \\ \dfrac{x+2}{x+1}, & 0 < x \le 3 \\ \dfrac{-x+3}{x+1}, & x > 3. \end{cases}$$

Ex. 4.74. a. $f : X \to Y, f(x) = 6 - x$ No function from Y to X since $4 \notin X$.

b. No function from X to Y since $4 \notin Y$. No function from Y to X since $-1 \notin X$.

c. No function from X to Y or from Y to X.

d. $f : X \to Y, f(x) = x$. No function from Y to X since $0 \notin X$.

Ex. 4.75. a. $f(\{1,3\}) = \{a,e\}$.
    b. $f^{-1}(\{a,b,c\}) = \{2,3\}$.
    c. $f^{-1}(f(\{2,4\})) = \{1,2,4\}$.
    d. $f(f^{-1}(\{b,d,e\})) = \{e\}$.

Ex. 4.76. a. $f((-2,5]) = [3,7)$.
    b. $f((-2,-1) \cup (3,6)) = (4,7)$.
    c. $f^{-1}([0,2)) = \emptyset$.
    d. $f^{-1}([4,7)) = (-2,6)$.
    e. $f^{-1}(f((-1,3))) = f^{-1}([3,7)) = (-2,6)$.
    f. $f(f^{-1}([-1,5])) = f([0,4]) = [3,5]$.

Ex. 4.77. $g(x) = 2x + 1$.

Ex. 4.78. $g(x) = x^2 - 3x + 2$.

Ex. 4.79. a. g is not one-to-one since $g(-1) = g(1) = -1$.
    b. h is one-to-one.
    c. k is one-to-one.

Ex. 4.80. a. $A = (-\infty, 3)$ or $A = (3, \infty)$.
    b. $A = (0, \pi)$ or $A = (\pi, 2\pi)$.

Ex. 4.82. $f^{-1}(y) = e^{\arcsin y}$.

Ex. 4.84. There are six one-to-one functions: $f_1(1) = a, f_1(2) = b$; $f_2(1) = a, f_2(2) = c$; $f_3(1) = b, f_3(2) = a$; $f_4(1) = b, f_4(2) = c$; $f_5(1) = c, f_5(2) = a$; $f_6(1) = c, f_6(2) = b$.

Ex. 4.86. Indeed, both $\langle x, x \rangle$ and $\langle x, -x \rangle$ are in g.

Ex. 4.87. a. The inclusion $f(A \cap B) \subseteq f(A) \cap f(B)$ is true for any function. For the other inclusion, let $y \in f(A) \cap f(B)$. There are $a \in A$ and $b \in B$ with $f(a) = f(b) = y$. Since f is one-to-one, we get $a = b \in A \cap B$, so $y \in f(A \cap B)$.
    b. Let $y \in f(A')$. Then $y = f(x)$ where $x \in X \setminus A$. We want to show that $y = f(x) \in Y \setminus f(A)$. To seek a contradiction, assume $f(x) \in f(A)$. Then $f(x) = f(a)$ for $a \in A$. Since f is one-to-one, it follows that $x = a \in A$, which contradicts the fact that $x \in X \setminus A$.
    For $X = \{1,2\}, Y = \{a,b,c\}, f(1) = a, f(2) = c$ and $A = \{1\}$ we have $f(A') = \{c\}$ but $(f(A))' = \{b,c\}$.
    c. Let $y \in (f(A))'$. Then $y \neq f(a)$ for all $a \in A$. Since f is onto, there is $x \in X$ with $f(x) = y$. It must be that $x \in A'$, hence $y \in f(A')$.
    For $X = \{1,2,3\}, Y = \{a,b\}, f(1) = f(2) = a, f(3) = b$ and $A = \{1\}$ we have $(f(A))' = \{b\}$ but $f(A') = \{a,b\}$.

Ex. 4.88. Assume $f(x_1) = f(x_2)$. Then $g(f(x_1)) = g(f(x_2))$. Since $g \circ f$ is injective, we get $x_1 = x_2$, hence f is injective.
    Let $f : \{1,2\} \to \{a,b,c\}, f(1) = a, f(2) = c$ and let $g : \{a,b,c\} \to \{u,v\}, g(a) = g(c) = u, g(b) = v$. Then f is injective, but $g \circ f : \{1,2\} \to \{u,v\}, (g \circ f)(1) = g(a) = u, (g \circ f)(2) = g(c) = u$, hence $(g \circ f)$ is not injective.

Assume $f : X \to Y$, $g : Z \to X$ and let $y \in Y$. Since $f \circ g : Z \to Y$ is surjective, there is $z \in Z$ with $f(g(z)) = y$. We found $x = g(z) \in X$ with $f(x) = y$, so f is surjective.

Let $f : \{1,2,3\} \to \{a,b\}, f(1) = f(2) = a, f(3) = b$ and let $g : \{u,v\}, g(u) = 1, g(v) = 2$. Then f is surjective, but $f \circ g : \{u,v\} \to \{a,b\}, (f \circ g)(u) = f(1) = a, (f \circ g)(v) = f(2) = a$, hence $f \circ g$ is not surjective.

Ex. 4.89. For example, $f \mid_{(-\infty,3/2)}$ and $f \mid_{(3/2,\infty)}$ are injective.

Ex. 4.90. 1. $f^{-1} : \mathbb{R} \to \mathbb{R}, f^{-1}(y) = \dfrac{y-1}{7}$.

2. $g^{-1} : [0,\infty) \to (-\infty,0], g^{-1}(y) = -\sqrt{y}$.

3. $h^{-1} : (-\infty,0] \to [2,\infty), h^{-1}(y) = 2 + \sqrt{-y}$.

4. $k^{-1} : [-6,3] \to [-3,1], k^{-1}(y) = \begin{cases} \frac{y}{3}, & 0 \le y \le 3 \\ \frac{y}{2}, & -6 \le y \le 0. \end{cases}$

5. $t^{-1} : [-1,1] \to [0,\pi/2], t^{-1}(y) = \dfrac{\pi}{4} + \arcsin \dfrac{y}{\sqrt{2}}$.

Ex. 4.91. a. $f_1 : (-\infty,3] \to [-7,\infty), f_1(x) = x^2 - 6x + 2$ has inverse

$$f_1^{-1} : [-7,\infty) \to (-\infty,3], f_1^{-1}(y) = 3 - \sqrt{y+7}.$$

b. $f_2 : [3,\infty) \to [-7,\infty), f_2(x) = x^2 - 6x + 2$ has inverse

$$f_2^{-1} : [-7,\infty) \to [3,\infty), f_2^{-1}(y) = 3 + \sqrt{y+7}.$$

c. $f_3 : (-\infty,0] \cup [3,6) \to [-7,\infty), f_3(x) = x^2 - 6x + 2$ has inverse

$$f_3^{-1} : [-7,\infty) \to (-\infty,0] \cup [3,6), f_3^{-1}(y) = \begin{cases} 3 - \sqrt{y+7}, & y \ge 2 \\ 3 + \sqrt{y+7}, & -7 \le y \le 2. \end{cases}$$

Ex. 4.92. $f \circ g \circ h : [1,\infty) \to [0,\infty), (f \circ g \circ h)(x) = \dfrac{4x-4}{x^2}$ has range $[0,1]$.

Ex. 4.94. $f(x) = \dfrac{-1 \pm |2x-3|}{2}$.

Ex. 4.96. Indeed, $x \in \limsup E_n$ if and only if for any $k \ge 1$ there is $n \ge k$ with $x \in E_n$. Also, $x \in \liminf E_n$ if and only if there is $k \ge 1$ such that $x \in E_n$ for all $n \ge k$.

Ex. 4.97. a) $\limsup E_n = \liminf E_n = (0,\infty)$.
   b) $\limsup E_n = \mathbb{Z}, \liminf E_n = \{1\}$.

Ex. 5.75. $R = \{\langle 2,2 \rangle, \langle 2,6 \rangle, \langle 2,12 \rangle, \langle 3,6 \rangle, \langle 3,12 \rangle, \langle 4,12 \rangle\}$,
$R^{-1} = \{\langle 2,2 \rangle, \langle 6,2 \rangle, \langle 12,2 \rangle, \langle 6,3 \rangle, \langle 12,3 \rangle, \langle 12,4 \rangle\}$,
$\mathrm{dom}(R) = \{2,3,4\}, \mathrm{ran}(R) = \{2,6,12\}$.

Ex. 5.76. $R = \{\langle 1, 3 \rangle, \langle 2, 4 \rangle, \langle 3, 1 \rangle, \langle 4, 2 \rangle\}$, $R \circ R = \{\langle 1, 1 \rangle, \langle 2, 2 \rangle, \langle 3, 3 \rangle, \langle 4, 4 \rangle\}$.

Ex. 5.78. a. $\text{dom}(R_1) = (-\infty, 3]$, $\text{ran}(R_1) = \mathbb{R}$.

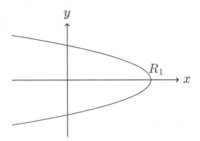

b. $\text{dom}(R_2) = [0, 1]$, $\text{ran}(R_2) = [-1, 1]$.

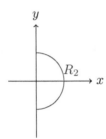

d. $\text{dom}(R_4) = \{-3\}$, $\text{ran}(R_4) = (-4, 4)$.

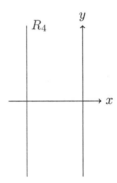

f. $\text{dom}(R_6) = [-9, 9]$, $\text{ran}(R_6) = [-9, 9]$.

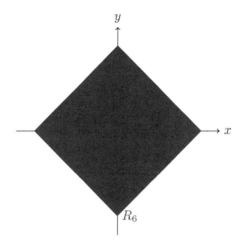

R_6

Ex. 5.79. a. $R^{-1} = \{\langle 1, a\rangle, \langle b, 2\rangle, \langle 4, 3\rangle, \langle y, x\rangle\}$.
  b. $S^{-1} = S$.
  c. $T^{-1} = \{\langle x, y\rangle \in \mathbb{R} \times \mathbb{R} : 3y^2 - 4x^2 = 9\}$.
  d. $V^{-1} = \{\langle x, y\rangle \in \mathbb{R} \times \mathbb{R} : x < 2y - 5\}$.
  e. $W^{-1} = \{\langle x, y\rangle \in \mathbb{R} \times \mathbb{R} : x(y + 3) = y\}$.

Ex. 5.81. a. $S \circ R = \{\{\langle x, y\rangle \in \mathbb{R} \times \mathbb{R} : 2(2x - 1)^2 + 3y^2 = 5\}$.
  b. $S \circ R = \{\langle x, y\rangle \in \mathbb{R} \times \mathbb{R} : y = \sin x^2\}$.

Ex. 5.83. a) R is transitive. Indeed, if $R \circ R \subseteq R$, then for $\langle x, z\rangle, \langle z, y\rangle \in R$ we have $\langle x, y\rangle \in R \circ R$, hence R is transitive. Assuming R transitive, given $\langle x, y\rangle \in R \circ R$, there is z with $\langle x, z\rangle, \langle z, y\rangle \in R$ and we get $\langle x, y\rangle \in R$.
  b) R' is symmetric.

Ex. 5.84. a) R is reflexive, symmetric and transitive.
  b) R is not reflexive, is symmetric and is not transitive.
  c) R is reflexive and symmetric but not transitive.

Ex. 5.86. a) $\mathcal{R}$ is only symmetric.
  b) $\mathcal{R}$ is only symmetric.
  c) $\mathcal{R}$ is reflexive, symmetric and transitive.
  d) $\mathcal{R}$ is reflexive and transitive.
  e) $\mathcal{R}$ is reflexive, symmetric and transitive.

Ex.5.88. Let $R = S = \{\langle 1, 3\rangle, \langle 2, 2\rangle, \langle 3, 1\rangle\}$.

Ex. 5.90. $R = id_X$ where $id_X : X \to X, id_X(x) = x$.

Ex. 5.91. R is reflexive since for all x we have $4 \mid x + 3x = 4x$. If xRy, then yRx since $y + 3x = 4(x + y) - (x + 3y)$, hence R is symmetric. If xRy and yRz, then xRz since $x + 3z = (x + 3y) + (y + 3z) - 4y$, hence R is transitive. The quotient set $\mathbb{Z}/R$ has four elements: $[0], [1], [2], [3]$.

Ex. 5.92. $R_g = \{\langle x, y \rangle \in [0, 2\pi] \times [0, 2\pi] : x + y = \pi \text{ or } x + y = 3\pi\}$. The quotient set $[0, 2\pi]/R_g$ can be identified with $[\pi/2, 3\pi/2]$.

Ex. 5.93. Indeed, if $[x]_4 = [y]_4$, then $4 \mid x - y$, hence $2 \mid x - y$ and $[x]_2 = [y]_2$.

Ex. 5.94. We have $[1]_3 = [4]_3$ but $[1]_5 \neq [4]_5$.

Ex. 5.95. Indeed, if $[x] = [y]$, then $5 \mid x - y$ and $5 \mid (2x + 3) - (2y + 3)$, hence $[2x + 3] = [2y + 3]$. The function f is one-to-one because $[2x + 3] = [2y + 3]$ implies $[x] = [y]$. It is also onto since the equation $[2x + 3] = [y]$ has solutions for all $[y] \in \mathbb{Z}_5$.

Ex. 5.97. The relation is reflexive since $f' = f'$ for all $f \in D$. It is symmetric since $f' = g'$ implies $g' = f'$. It is transitive since $f' = g'$ and $g' = h'$ implies $f' = h'$. We have $f \sim g$ iff $f - g$ is constant.

$$[f] = \{g \in D : g = f + c, c \in \mathbb{R}\}.$$

Ex. 5.100. X/R is obtained from a semicircle with the endpoints identified.

Ex. 5.101. Let xSy and ySz. Then xRy, yRz and $x \neq y, y \neq z$. Since R is transitive, we get xRz. If $x = z$, then since R is antisymmetric, we get $x = y$, a contradiction. It follows that xSz, so S is transitive.

Ex. 5.102. a) Let $X = \{2, 3, 6\}$, let $S = \mid$ and let $R = S^{-1}$. Then $\langle 2, 3 \rangle \in R \circ S$ because $\langle 2, 6 \rangle \in S, \langle 6, 3 \rangle \in R$. Also $\langle 3, 2 \rangle \in R \circ S$ but $2 \neq 3$, so $R \circ S$ is not antisymmetric.

b) Let $X = \{1, 2, 3\}$, let $R = \{\langle 1, 1 \rangle, \langle 1, 2 \rangle, \langle 2, 2 \rangle, \langle 3, 3 \rangle\}$ and let $S = \{\langle 1, 1 \rangle, \langle 2, 3 \rangle, \langle 2, 2 \rangle, \langle 3, 3 \rangle\}$, which are partial orders on X, but $R \cup S = \{\langle 1, 1 \rangle, \langle 1, 2 \rangle, \langle 2, 3 \rangle, \langle 2, 2 \rangle, \langle 3, 3 \rangle\}$ is not transitive, so $R \cup S$ is not a partial order on X.

Ex. 5.103. R is not antisymmetric.

Ex. 5.104. There are six total orders.

Ex. 5.106. The minimal elements are $\langle 0, n \rangle$ and $\langle n, 0 \rangle$ where $n \in \mathbb{N}$. For example, let $A = \{\langle 1, 2 \rangle, \langle 2, 1 \rangle\}$.

Ex. 5.107. $(Y, \mid)$ is not a lattice since $2 \vee 3 = 6$ is not in Y.

Ex. 5.108. We have $\langle x_1, x_2 \rangle \vee \langle y_1, y_2 \rangle = \langle \max(x_1, y_1), \max(x_2, y_2) \rangle$ and $\langle x_1, x_2 \rangle \wedge \langle y_1, y_2 \rangle = \langle \min(x_1, y_1), \min(x_2, y_2) \rangle$. The elements $\langle 1, 2 \rangle, \langle 2, 1 \rangle$ are incomparable, hence $\preceq$ is not a total order.

The dictionary order $\preceq_L$ is a total order, since given $\langle x_1, x_2 \rangle, \langle y_1, y_2 \rangle \in \mathbb{Z} \times \mathbb{Z}$ or $\mathbb{R} \times \mathbb{R}$ we have $\langle x_1, x_2 \rangle \preceq_L \langle y_1, y_2 \rangle$ or $\langle y_1, y_2 \rangle \preceq_L \langle x_1, x_2 \rangle$. In particular, $\langle 1, 2 \rangle \preceq_L \langle 2, 1 \rangle$.

Ex. 5.109. Let $A \subseteq \mathbb{N} \times \mathbb{N}$ be nonempty. Since $\mathbb{N}$ with usual order is well-ordered, the set of first components of the elements of A has a smallest element a. Similarly, the set of second components has a smallest element b. The element $\langle a, b \rangle$ has the property that $\langle a, b \rangle \preceq_L \langle x, y \rangle$ for all $\langle x, y \rangle \in A$.

Ex. 5.110. The set $\{\langle n, n\rangle : n \leq 0\}$ has no smallest element.

Ex. 6.41. a) Assuming $x \leq y$, there is $u \geq 0$ with $y = x + u$. It follows that $y + z = x + z + u$, hence $x + z \leq y + z$. The other implication is proved using cancelation.

b) It follows from the law of trichotomy.

Ex. 6.42. Let $s : E \rightarrow E, s(n) = n + 2$.

Ex. 6.43. Notice that $D(n) = \{n, n+1, n+2, ...\}$, $D(s(n)) = \{n+1, n+2, ...\} = s(D(n))$. For 5, let k be the smallest element of A.

Ex. 6.44. a) $x^{y+z} = \underbrace{x \cdots x}_{y+z} = \underbrace{x \cdots x}_{y} \cdot \underbrace{x \cdots x}_{z}$ or use induction on z.

b) For $z = 0$ we have $x^{y \cdot 0} = 1 = (x^y)^0$. Assume $x^{yk} = (x^y)^k$. Then

$$x^{y(k+1)} = x^{yk+y} = (x^y)^k \cdot x^y = (x^y)^{k+1}.$$

Ex. 6.45. We have $u_0 = 2^0 + 1 = 2$. Assume $u_j = 2^j + 1$ for $0 \leq j \leq k$. Then

$$u_{k+1} = 3u_k - 2u_{k-1} = 3(2^k + 1) - 2(2^{k-1} + 1) = 2^{k+1} + 1.$$

Ex. 6.46. This is true for $n = 1 = 1 \cdot 2^0$. Assume that any positive integer $n \leq k$ is a product of an odd integer and a power of 2. Consider $k + 1$. If this is odd, we are done. If $k + 1$ is even, then $k + 1 = 2 \cdot m$, where $m \leq k$. Let $m = p \cdot 2^q$ with p odd. Then $k + 1 = p \cdot 2^{q+1}$.

Ex. 6.47. For $n = 3$ there is only one subset and $3 \cdot 2 \cdot 1/3! = 1$. Assume this is true for $n = k$ and consider a set with $k+1$ elements $B = \{b_1, b_2, ..., b_k, b_{k+1}\}$. The subsets with three elements of B are of two kinds: subsets with three elements not containing b_{k+1} and subsets containing b_{k+1}. The total number of subsets is

$$\frac{k(k-1)(k-2)}{3!} + \frac{k(k-1)}{2!} = \frac{k(k-1)(k-2) + 3k(k-1)}{3!} = \frac{(k+1)k(k-1)}{3!}.$$

Ex. 6.48. We first show that this is true for $n = 8, 9, 10$. Indeed,

$$8 = 3 + 5, 9 = 3 \cdot 3, 10 = 2 \cdot 5.$$

Assuming $k = 3a + 5b$ with $a, b \geq 0$, we have $k + 3 = 3(a + 1) + 5b$.

Ex. 7.34. b. We have

$$|x| = |x - y + y| < |x - y| + |y|, \quad |y| = |y - x + x| \leq |y - x| + |x|,$$

hence $|x| - |y| \leq |x - y|$ and $|x| - |y| \geq -|x - y|$, so $||x| - |y|| \leq |x - y|$.

Ex. 7.36. e) Let $x = zu, y = yv, z = tw$. Then $x = tuw, y = tvw$, so $x/t = uw, y/t = vw$.

Ex. 7.38. We may assume $n \geq 2$. If $p \leq x$ for any prime, then there are only finitely many primes, a contradiction with Example 2.13.

Ex. 7.40. This is false: $4 \mid 2 + 6$ but $4 \nmid 2$ and $4 \nmid 6$.

Ex. 7.42. If $d \mid n$ and $d \mid n + 6$, it follows that $d \mid (n + 6) - n = 6$, so $d \in \{1, 2, 3, 6\}$.

Ex. 7.43. This is true for $n = 1$. Let $u, v \in \mathbb{Z}$ with $1 = au + bv$. Assuming $\gcd(a, b^k) = 1$, let $r, s \in \mathbb{Z}$ with $1 = ar + b^k s$. Multiplying the two equalities, we get $1 = a(aru + ub^k s + rbv) + b^{k+1} vs$, so $\gcd(a, b^{k+1}) = 1$.

Ex. 7.44. Consider $S = \{ax + by + cz : x, y, z \in \mathbb{Z}\}$. The set $S \cap \mathbb{P}$ is not empty and it has a smallest element d. Using the same method as in the proof of Theorem 7.11 we get $d = \gcd(a, b, c)$ and $d = as + bt + cu$.

Ex. 7.45. By completing the square, $x^4 = (2y+1)^2 - 24$, so $(2y+1)^2 - x^4 = 24$. Factoring, we get $(2y + 1 - x^2)(2y + 1 + x^2) = 24$. Considering the divisors of 24, the only systems with integer solutions are

$$2y + 1 - x^2 = 4, \quad 2y + 1 + x^2 = 6$$

and

$$2y + 1 - x^2 = -6, \quad 2y + 1 + x^2 = -4$$

with solutions $x = \pm 1, y = 2$ and $x = \pm 1, y = -3$.

Ex. 7.47. Replace $10 \equiv 1 \pmod 9$ with $10 \equiv 1 \pmod 3$ in the proof of Theorem 7.30.

Ex. 7.48. Let $n = 2k + 1$. Then $n^2 - 1 = (2k + 1)^2 - 1 = 4k^2 + 4k = 4k(k + 1)$ is divisible by 8 since for any integer k, the product $k(k + 1)$ is even.

Ex. 7.49. We have $n^3 - n = n(n-1)(n+1)$. Now n divided by 3 gives remainders $0, 1$ or 2, so $n = 3k$ or $n = 3k + 1$ or $n = 3k + 2$. In any case, the product $n(n - 1)(n + 1)$ is a multiple of 3. Since $n(n - 1)$ is always even, it follows that $n^3 - n$ is a multiple of $2 \cdot 3 = 6$.

Ex. 7.50. 1. Considering the remainders $0, 1, 2, 3, 4, 5, 6$ of x when divided by 7, we see that $x = 2$ is a solution.

2. Considering the remainders $0, 1, 2, 3, 4$ of x when divided by 5, there is no solution since x^2 is congruent with $0, 1$ or 4 modulo 5.

Ex. 7.52. Suppose that $2n + 1 = x^2$ and $3n + 1 = y^2$. Then $x^2 + y^2 \equiv 2 \pmod 5$. But $x^2 \equiv 0, 1, 4 \bmod 5$, so we must have $x^2 \equiv y^2 \equiv 1 \pmod 5$. Therefore $5 \mid n$. On the other hand, since x must be odd, n must be even. Then y must be odd. Let $x = 2a + 1$ and $y = 2b + 1$. Now notice that $n = y^2 - x^2 = 4(b^2 + b - a^2 - a)$. The number in parentheses must be even. Hence $8 \mid n$. Hence $40 \mid n$.

Ex. 8.30. Let $|A| = n$ and $f : A \to A$ an onto function, so $f(A) = A$. If f is not one-to-one, then there are $a_1 \neq a_2$ with $f(a_1) = f(a_2)$, so $|f(A)| \leq n - 1$, a contradiction. Conversely, if $f : A \to A$ is one-to-one, then $|f(A)| = n$, so f must be onto.

Ex. 8.31. Using the axiom of choice, we can find a countable subset $B = \{a_1, a_2, a_3, ...\}$ of A. Consider $f : B \to B, f(a_n) = a_{n+1}$ and extend f to $\tilde{f} : A \to A$ such that $\tilde{f}(x) = x$ for $x \in A \setminus B$. Then $\tilde{f}$ is one-to-one, but not onto.

Ex. 8.32. Consider the function $F : \mathcal{P}(A) \to \{0, 1\}^A, F(B) = \chi_B$, where

$$\chi_B : A \to \{0, 1\}, \chi_B(x) = \begin{cases} 1 & \text{if } x \in B \\ 0 & \text{if } x \in A \setminus B \end{cases}$$

is the characteristic function of B. Then F is one-to-one since $\chi_B = \chi_C$ implies $B = C$ and onto since given $f : A \to \{0, 1\}$, by taking $D = f^{-1}(1)$ we have $f = \chi_D$.

Ex. 8.33. Notice that R is reflexive, antisymmetric and transitive, and so is an order relation. Let $A \subseteq \mathbb{Q}$ be not empty. Then $\phi(A) \subseteq \mathbb{N}$ is not empty, so it has a smallest element n. Then $a = \phi^{-1}(n)$ has the property that $\phi(a) \leq \phi(x)$ for all $x \in A$, hence $(\mathbb{Q}, R)$ is a well-ordered set.

Ex. 8.34. Consider the countable sets $A_n = \{a_{n1}, a_{n2}, ...\}$ for $n \geq 1$. To show that $A = \bigcup_{n=1}^{\infty} A_n$ is countable, we define a bijection $f : \mathbb{P} \to A$ using the following pattern

$$f(1) = a_{11}, f(2) = a_{12}, f(3) = a_{21}, f(4) = a_{31},$$

$$f(5) = a_{22}, f(6) = a_{13}, f(7) = a_{14}, f(8) = a_{23}, ...$$

Ex. 8.35. We know that $\mathbb{N}^2 = \mathbb{N} \times \mathbb{N}$ is countable. By induction it is easy to prove that $\mathbb{N}^n$ is countable for any $n \geq 1$. Let S_n be the set of subsets of $\mathbb{N}$ with n elements. We define $f_n : S_n \to \mathbb{N}^n$ by $f(\{a_1, a_2, ..., a_n\}) = \langle a_1, a_2, ..., a_n \rangle$ where $a_1 < a_2 < \cdots < a_n$. Then f_n is injective, hence S_n is countable. Then $\mathcal{P}_f(\mathbb{N}) = \{\emptyset\} \cup \bigcup_{n=1}^{\infty} S_n$ is a countable union of countable sets, and hence is countable.

Ex. 8.36. If $\mathbb{R} \setminus \mathbb{Q}$ is countable, then $\mathbb{R} = (\mathbb{R} \setminus \mathbb{Q}) \cup \mathbb{Q}$ would be countable, a contradiction.

Ex. 8.37. Suppose we can find a bijection $f : \mathbb{N} \to S$. Construct an element $s \in S$ such that

$$s_n = \begin{cases} 0 & \text{if } f(n)_n = 1 \\ 1 & \text{if } f(n)_n = 0 \end{cases}.$$

Then $f(n) \neq s$ for all $n \in \mathbb{N}$, a contradiction.

Ex. 8.39. We have $|A| \leq |B| \leq |C| \leq |A|$, so $A \approx B \approx C$.

Ex. 9.36. $100 + 100 + 100 - 50 - 50 - 50 + 25 = 175$.

Ex. 9.39. a) There are $\lfloor 5000/3 \rfloor = 1666$ integers divisible by 3, $\lfloor 5000/4 \rfloor = 1250$ divisible by 4, and 416 divisible by 12. The required number is $5000 - 1666 - 1250 + 416 = 2500$.

   b) $\lfloor 5000/4 \rfloor + \lfloor 5000/5 \rfloor + \lfloor 5000/6 \rfloor - \lfloor 5000/20 \rfloor - \lfloor 5000/24 \rfloor - \lfloor 5000/30 \rfloor + \lfloor 5000/120 \rfloor = 1250 + 1000 + 833 - 250 - 208 - 166 + 41 = 2500$.

Ex. 9.40. $P(26,3)P(10,3) = 11,232,000$ license plates.

Ex. 9.41. There are ten digits, but we exclude the 0 digit at the beginning. We get $P(10,3) - P(9,2) + P(10,2) - P(9,1) + P(10,1) - 1 = 738$ integers with distinct digits.

Ex. 9.42. Consider a grid which divides the square into $7 \cdot 7 = 49$ little squares. By the pigeonhole principle, three points must be in the same little square. Since the diameter of the circumscribed circle is $\sqrt{2}/7$, a disc of radius $1/7$ with the same center as the circumscribed circle will contain the little square in its interior.

Ex. 9.43. By the pigeonhole principle, two of the integers must have difference ± 1.

Ex. 9.45. No.

Ex. 9.51. $a = 2^{n+1} - 1$.

Ex. 9.53. In the identity $(1+x)^{m+n} = (1+x)^m (1+x)^n$, identify the coefficients of x^k.

Ex. 9.55. a) $a_n = \dfrac{1}{3}(5 \cdot 2^n + 4(-1)^n)$.
   b) $a_n = 2^{n/2-1}(\sqrt{2} + 1 + (-1)^n(1 - \sqrt{2}))$.
   c) $a_n = 3 - 2^{1-n}$.
   d) $a_n = 2^{n+1} - \frac{n+2}{2}$.

Ex. 9.56. $a_n = 2^{2n+1}$.

Ex. 9.57. The recurrence is $a_3 = 6, a_n = a_{n-1} + n, n \geq 4$.

Ex. 9.58. We have $b_1 = 2, b_2 = 3, b_3 = 5$. The recurrence relation is $b_n = b_{n-1} + b_{n-2}$ for $n \geq 3$. We get a translation of the Fibonacci sequence.

Ex. 10.42. a. $6 - 12 = 8 - 14 = -6$.
   b. $a - b = c - c = 0 \Rightarrow a = b$.
   c. $c - c - 5 = 4 - b \Rightarrow b = 9$.

Ex. 10.43. a. From $\langle a,b \rangle \sim \langle a',b' \rangle$ and $\langle c,d \rangle \sim \langle c',d' \rangle$ we get $a + b' = a' + b$ and $c + d' = d + c'$. By adding these together, we obtain $a + d + b' + c' = a' + d' + b + c$. Since $a + d < b + c$, there is $u > 0$ with $b + c = a + d + u$. We get $a + d + b' + c' = a' + d' + a + d + u$ and by cancellation, $b' + c' = a' + d' + u$, so $a' + d' < b' + c'$.

b. Not well-defined since $[0,1] < [2,1]$ since for example $0 < 2$ but $[0,1] = [3,4]$ and $3 > 2$.

Ex. 10.45. a. No; b. No; c. Yes; d. Yes; e. No; f. Yes; g. Yes.

Ex. 10.47. Not well-defined: $[0,1] = [0,2]$ but $0 - 1 \neq 0 - 2$.

Ex. 10.48. Let $a < b$ be two rational numbers. Then $a < a + \dfrac{b-a}{n} < b$ for all $n \geq 2$.

Ex. 10.52. $\dfrac{4}{21} = \dfrac{1}{9} + \dfrac{6}{9^2} + \dfrac{3}{9^3} + \dfrac{7}{9^4} + \dfrac{6}{9^5} + \dfrac{3}{9^6} + \dfrac{7}{9^7} + \cdots = 0.1\overline{637}$ in base 9.
Notice that the period starts later since $21 = 3 \cdot 7$ has a factor of 3.

Ex. 11.79. Indeed, $1 \in C, 0 < 1$ but $0 \notin C$.

Ex. 11.80. a. By definition, C is positive if there is $r > 0$ with $r \in C$. It follows that $0 \in C$. Conversely, if $0 \in C$, then there is $r > 0$ with $r \in C$.

b. There is $r > 0$ with $r \in C \setminus D$ and there is $s < 0$ with $s \notin D$. Since for all $x \in D$ we have $x < s < r$, it follows that $D \subset C$, i.e., $D \prec C$.

c. Let $r < 0$ with $r \notin C$. Then $r \notin D$ and D is negative.

d. If $C \prec \hat{0}$, there is $r \in \hat{0}$ such that $r \notin C$. It follows that C is negative. Conversely, if C is negative, then there is $r < 0$ with $r \notin C$. It follows that $x < r$ for all $x \in C$, hence $C \prec \hat{0}$.

Ex. 11.81. For $r = 0$ or $s = 0$ the equality is obvious. Assume first that $r, s > 0$ and let $x \in \hat{r}\hat{s}$. By definition there are $0 < c < r$ and $0 < d < s$ with $x \leq c \cdot d$. It follows that $x < rs$, hence $x \in \widehat{rs}$. For the other inclusion, given $x \in \widehat{rs}$ we must find $0 < a < r$ and $0 < b < s$ with $x \leq a \cdot b$. Since $x < rs$, we get $x/s < r$. Let $0 < a < r$ with $x/s \leq a$. Since $x/a < s$, we can find $0 < b < s$ with $x/a < b$. We get $x \leq a \cdot b$.

The other cases can be reduced to the previous case.

Ex. 11.82. Since $x \cdot x < 2$ for $x > 0$, we get $D \cdot D \preceq \hat{2}$. For the other inequality, let $x \in \hat{2}$. It suffices to take $0 < x < 2$. If $x^2 < 2$, we can write $x = x \cdot 1$ with $x, 1 \in D$. Assume $x^2 > 2$. Choose $y, z \in D$ such that $x \leq yz$. This shows that $x \in D \cdot D$, hence $\hat{2} \preceq D \cdot D$.

Ex. 11.85. If $x = a/b$ with $x^3 = 7$, we get $a^3 = 7b^3$. We may assume that $\gcd(a,b) = 1$. Since $7 \mid a^3$ it follows that $7 \mid a$. Replacing a by $7c$ we get $b^3 = 49c^3$, in particular $7 \mid b$, a contradiction.

The real number denoted $\sqrt[3]{7}$ is given by the cut $C = \{x \in \mathbb{Q} : x^3 < 7\}$.

Ex. 11.88. Since $(\sqrt{3}+\sqrt{2})^2 = 5+2\sqrt{6}$ it follows that $\sqrt{2}+\sqrt{3}-\sqrt{5+2\sqrt{6}} = 0$.

Ex. 11.89. 1. If $L_1 \neq L_2$ are two limits, let $\varepsilon < |L_1 - L_2|/2$. By the definition of limit, there is $N_1 \geq 1$ such that $|x_n - L_1| < \varepsilon$ for all $n \geq N_1$. Also, there is $N_2 \geq 1$ such that $|x_n - L_2| < \varepsilon$ for all $n \geq N_2$. For $n \geq N = \max(N_1, N_2)$, we have $x_n \in (L_1 - \varepsilon, L_1 + \varepsilon) \cap (L_2 - \varepsilon, L_2 + \varepsilon) = \emptyset$, a contradiction.

2. Let $x_n \to L$. Given $\varepsilon > 0$, there is $N \geq 1$ such that $|x_n - L| < \varepsilon/2$ for all $n \geq N$. For $m, n \geq N$ we get

$$|x_n - x_m| \leq |x_n - L| + |x_m - L| < \varepsilon/2 + \varepsilon/2 = \varepsilon,$$

hence (x_n) is Cauchy.

Ex. 11.90. The sequences are partial sums of convergent series, and therefore are convergent and Cauchy.

Ex. 11.92. The number $1 - \sqrt{3}$ is a root of $x^2 - 2x - 2 = 0$. The number $\sqrt{5} - \sqrt{2}$ is a root of $x^4 - 14x^2 + 9 = 0$.

Ex. 11.93. Assume α is a root of $a_n x^n + a_{n-1} x^{n-1} + \cdots + a_1 x + a_0 = 0$. Then $k\alpha$ is a root of the polynomial $a_n x^n + k a_{n-1} x^{n-1} + \cdots + k^{n-1} a_1 + k^n a_0 = 0$. If $\alpha \neq 0$, then $1/\alpha$ is a root of $a_n + a_{n-1} x + \cdots + a_1 x^{n-1} + a_0 x^n = 0$.

Ex. 11.95. We have $i^{4k} = 1, i^{4k+1} = i, i^{4k+2} = -1$ and $i^{4k+3} = -i$.

Ex. 11.98. Each polynomial $p(x) = a_n x^n + a_{n-1} x^{n-1} + \cdots + a_1 x + a_0$ with $a_i \in \mathbb{Z}$ determines an element in $\mathbb{Z}^{n+1}, n \geq 1$. Since $\mathbb{Z}^{n+1}$ is countable, there are countably many equations $p(x) = 0$, each having $\deg p$ roots.

Ex. 11.99. The midpoints of the sides are $\dfrac{z_1 + z_2}{2}, \dfrac{z_2 + z_3}{2}, \dfrac{z_3 + z_1}{2}$. The medians intersect at a point dividing one median in the ratio $2 : 1$. Its coordinate is $\dfrac{z_1}{3} + \dfrac{2}{3} \cdot \dfrac{z_2 + z_3}{2} = \dfrac{1}{3}(z_1 + z_2 + z_3)$.

Ex. 11.100. One square has the other two vertices $i(w - z) + z$ and $i(w - z) + w$. Another square has $-i(w - z) + z$ and $-i(w - z) + w$. For the last one, z and w determine a diagonal, so the other vertices are $\frac{1}{2}(i(w - z) + z + w)$ and $\frac{1}{2}(-i(w - z) + z + w)$.

Ex. 11.101. a) $x_1 = 2i, x_2 = -i$. b) $x_1 = \dfrac{1 - \sqrt{3}}{2}(1 + i), x_2 = \dfrac{1 + \sqrt{3}}{2}(1 + i)$.

Ex. 11.102. The set $A = \{ni : n \in \mathbb{Z}\}$ is bounded above by 1 since $ni \prec 1$ for all n, but there is no least upper bound.

Ex. 11.104. Write $z = \cos t + i \sin t$. Then $1 + z = 2 \cos \dfrac{t}{2}(\cos \dfrac{t}{2} + i \sin \dfrac{t}{2})$ and $1 - z = 2 \sin \dfrac{t}{2}(\sin \dfrac{t}{2} + i \cos \dfrac{t}{2})$. It follows that $|1 + z|^2 + |1 - z|^2 = 4$.

Ex. 11.105. We have $w = 2(\cos \dfrac{7\pi}{4} + i \sin \dfrac{7\pi}{4})$. The roots are $z_k = \sqrt[4]{2}(\cos \dfrac{2k\pi + 7\pi/4}{4} + i \sin \dfrac{2k\pi + 7\pi/4}{4})$ for $k = 0, 1, 2, 3$.

Ex. 11.106. Let $\omega = \cos \frac{\pi}{3} + i \sin \frac{\pi}{3}$. Then $z_3 = (z_2 - z_1)\omega + z_1$ and $\omega^2 + \omega + 1 = 0$. A tedious computation shows that $z_1^2 + z_2^2 + z_3^2 = z_1 z_2 + z_1 z_3 + z_2 z_3$.

Ex. 11.107. The equation $z^n - 1 = 0$ can be factored as

$$(z - 1)(z^{n-1} + z^{n-2} + \cdots + z + 1) = 0.$$

We may assume $z_0 = 1$ and $z_1 = \cos \frac{2\pi}{n} + i \sin \frac{2\pi}{n}$. Then $z_k = z_1^k$ for $2 \leq k \leq n - 1$ and hence $z_0 + z_1 + \cdots + z_{n-1} = 0$. One can also use the relations between roots and coefficients.

Ex. 11.108. Since $z + \dfrac{1}{z} = 1$, we get $z^2 - z + 1 = 0$ and $z = \cos \dfrac{\pi}{3} \pm \sin \dfrac{\pi}{3}$. It follows that $z^n + \dfrac{1}{z^n} = 2 \cos \dfrac{n\pi}{3}$.

Ex. 11.109. Since $1 - i = \sqrt{2}(\cos \frac{\pi}{4} - i \sin \frac{\pi}{4})$, using De Moivre's formula we get

$$(1 - i)^{-20} = (\sqrt{2})^{-20}(\cos \frac{-20\pi}{4} - i \sin \frac{-20\pi}{4}) = -2^{-10}.$$

Similarly,

$$1 + i = \sqrt{2}(\cos \frac{\pi}{4} + i \sin \frac{\pi}{4}),$$

so

$$(1 + i)^{75} = (\sqrt{2})^{75}(\cos \frac{75\pi}{4} + i \sin \frac{75\pi}{4}) = 2^{37}(-1 + i).$$

Ex. 11.110. Using the binomial formula, $(\cos t + i \sin t)^3 = \cos^3 t + 3i \cos^2 t \sin t - 3 \cos t \sin^2 t - i \sin^3 t$. On the other hand, $(\cos t + i \sin t)^3 = \cos 3t + i \sin 3t$. Identifying the real and imaginary parts, we get $\cos 3t = \cos^3 t - 3 \cos t \sin^2 t$ and $\sin 3t = 3 \cos^2 t \sin t - \sin^3 t$.

Ex. 11.111. Write $z = |z|(\cos t + i \sin t)$. Then $\dfrac{z}{|z|} + \dfrac{|z|}{z} = \cos t + i \sin t + \cos t - i \sin t = 2 \cos t$.

Ex. 11.113. We have $\langle z, w \rangle = \langle z, 0 \rangle \langle 0, 1 \rangle + \langle w, 0 \rangle \langle 0, i \rangle = z + wj$.

Index

absolute value, 17, 105, 149, 160, 167
addition of positive integers, 92
algebraic number, 172
algebraically closed field, 172
antecedent, 4
antisymmetric, 81
Archimedean property, 95
argument of z, 175
at most countable set, 120
axiom, 15, 16
axiom of choice, 68

base, 100
biconditional, 6
bijective, 63
binomial formula, 131

cancelation law, 94
Cantor's diagonal argument, 123
Cantor–Bernstein theorem, 119
cardinality, 117
Cartesian product, 46
Cartesian product of a family of sets, 67
Catalan numbers, 137
Cauchy sequence, 163, 168
Cayley number, 177
ceiling function, 54
chain, 85
characteristic equation, 133
characteristic function, 55
codomain, 52
coextension, 59
combination, 129
common divisor, 107
common multiple, 107
comparable elements, 81

complement, 41
complete induction, 26
complete ordered field, 161
complex number, 172
composition, 62, 77
conclusion, 4
conditional, 4
congruence modulo n, 79
conjecture, 16
conjunction, 3
consequent, 4
constant function, 53
continuum, 123
continuum hypothesis, 123
contradiction, 8
contrapositive, 7
contrapositive proof, 19
converse, 7
corestriction, 59
corollary, 16
countable set, 120
counterexample, 11, 25

De Moivre formula, 176
De Morgan's laws, 8, 42
decimal fraction, 149
Dedekind cut, 154
descendent, 103
diagonal, 53
dictionary order, 90
Diophantine equation, 111
direct image, 57
direct proof, 17
directed set, 86
disjoint, 38
disjoint union, 47
disjunction, 3